MANNED AND UNMANNED FLIGHTS TO THE

MOON

This book is dedicated to the memory of those astronauts and cosmonauts
who gave their lives in the pursuit of space exploration.

If we die, we want people to accept it. We are in a risky business, and
we hope that if anything happens to us, it will not delay the program.
The conquest of space is worth the risk of life.

Gus Grissom – Astronaut
Mercury 7 [Liberty Bell], Gemini 3 and *Apollo 1.*

MANNED AND UNMANNED FLIGHTS TO THE
MOON

TERRY C. TREADWELL

WHITE OWL

AN IMPRINT OF PEN & SWORD BOOKS LTD.
YORKSHIRE – PHILADELPHIA

First published in Great Britain in 2024 by
White Owl
An imprint of
Pen & Sword Books Ltd
Yorkshire – Philadelphia

ISBN 978 1 39903 927 7

Typeset in 11/13 pts Cormorant Infant
by SJmagic DESIGN SERVICES, India.
Printed and bound in India by Parksons Graphics Pvt. Ltd.

Pen & Sword Books Ltd. incorporates the imprints of Pen & Sword Books: After the Battle, Archaeology, Atlas, Aviation, Battleground, Discovery, Family History, History, Maritime, Military, Politics, Select, Transport, True Crime, Fiction, Frontline Books, Leo Cooper, Praetorian Press, Seaforth Publishing, Wharncliffe and White Owl.

For a complete list of Pen & Sword titles please contact

PEN & SWORD BOOKS LIMITED
George House, Beevor Street, Off Pontefract Road, Hoyle Mill, Barnsley, South Yorkshire, England, S71 1HN
E-mail: enquiries@pen-and-sword.co.uk
Website: www.pen-and-sword.co.uk

or

PEN AND SWORD BOOKS
1950 Lawrence Rd, Havertown, PA 19083, USA
E-mail: uspen-and-sword@casematepublishers.com
Website: www.penandswordbooks.com

Contents

Introduction

This is the story of how the dreams of exploring other worlds became a reality with the vision of early scholars and scientists, culminating in the development of the rockets that launched man into space.

The exploration of space today has gone forward in leaps and bounds, with the planet Mars now 'inhabited' by unmanned rovers and drones, all being controlled from Earth. With the *Voyager* spacecraft now in deep space after leaving the solar system in 1977, and a number of other unmanned spacecraft being sent to examine the atmospheres of other far distant planets and their moons, such as Jupiter, Europa, Mercury, Venus and Pluto, we tend to forget the beginnings of how all this came about.

This book is about the American Mercury, Gemini, Apollo and Russian Soyuz space programmes, covering the early years of space development. The Russians were the first to make serious inroads into the use of rockets for space exploration and were the first to land an unmanned spacecraft on the Moon. However, in their desperate attempt to beat the Americans in the so called 'Space Race', they suffered a number of incidents and failures, some of which were fatal, which highlighted the dangers that all the astronauts and cosmonauts faced when safety was ignored in an effort to gain political 'one-upmanship'. Although the early spaceflights appeared to have nothing to do with the unmanned and manned missions to the Moon, they are included because these were the forerunners of these missions, the 'stepping stones', as everything was designed and planned to eventually place a man on the lunar surface.

The Ranger programme missions in the 1960s were launched to photograph the Moon's surface and were instrumental in helping to find landing sites for future manned flights. The exploration of the Moon by astronauts in July 1969, when man first set foot on another world, was a major step forward. With the successes of America and Russia in their lunar programmes, came other countries such as China, India, Japan and Saudi Arabia, all wanting to be part of this world of exploration.

There are now plans to send astronauts back to the Moon in 2026 and, sometime in the near future, to Mars.

CHAPTER ONE

In the Beginning

Since the beginning of time man has looked at the myriad of stars that carpeted the sky and wondered what secrets these bright twinkling specks held. Then later, with the aid of telescopes, he started to look at them more closely and tried to discover more about them.

The Greeks were the first recorded people to realise that the Moon was the nearest object to the Earth, and that if man were to take the first tentative steps in space, that is where he had to start. The first foundations of astronomy were by Greek scholars around 624 BC when the teachings of Thales of Miletus were recorded. They realised that the Earth was just a tiny speck in the universe and man was just an infinitesimal part of the whole thing. The first ideas of objects in space were obscure to say the least. The Persian [Iranian] Heraclitus of Ephesus [6th century BC] claimed that the Moon was in fact only one foot in diameter, whilst the Italian scholar Xenophanes of Colophon [478 BC] maintained that the earth was surrounded by many Suns and Moons, based upon regions and zones. This theory however was from a man who also thought the world was flat!

There were however some men who had other ideas and amongst these was Anaxagoras of Clazomenæ [500 BC to 428 BC] who was convinced that the Moon had mountains, ravines and plains. Over the next thousand years numerous philosophers, writers, scholars and scientists came up with stories about the Moon and other worlds and the ways of travelling to them. As the Greek and Roman empires collapsed, so did their great libraries, and almost all the information they contained disappeared. It was left to the Arab world to rekindle interest in the stars and here they drew up new charts that enabled people to plot their movement. One of these scholars was the Egyptian, Claudius Ptolomey [100–170 AD], who wrote an astronomical treatise known as *The Almagest*, where he claimed that the Earth was the centre of the known universe. It was this astronomical declaration that became widely accepted by the various religious organisations and scholars, remaining unchallenged for the next 1,000 years or so. It appears that almost all these scholars were from the Middle and Far East, and Baghdad became the astronomical centre of the known world. Europe however had slumped into the Dark Ages, when anyone even questioning or challenging anything to do with the stars and planets left themselves wide open to charges of heresy. It has to be remembered that in these early years very few people could read and write and so were at the mercy of those who could.

It wasn't until the sixteenth century that interest in the stars was re-awakened in Europe, when in 1543, Polish mathematician and astronomer Nicolaus Copernicus published his book *De Revolutionibus Orbium Cælestium* [Concerning the Revolutions of the Celestial Orbs]. Using the Ptolemaic system, which stated that the Sun, Moon and all other celestial bodies move round the earth, he changed the central body to be the Sun. Wisely he held back on the publication of his book until the last days of his life, because he knew that the Roman Catholic Church's doctrine at that time still claimed that the earth was the centre of the universe. The fact that Nicolaus Copernicus was also a canon in the Catholic Church at the time made it even more dangerous. Had he published his book some years earlier, there is no doubt he would have been arrested, tried and convicted of heresy and most likely executed. There is little doubt about this, as one of his followers, Giordano Bruno, was arrested, tried and burned at the stake in Rome in 1600 for his 'heretic' teachings using Copernicus's book. But despite these initial setbacks science progressed, and, with the invention of the telescope in 1608, went forward in leaps and bounds.

Galileo Galilei [1564–1642], the Italian scientist, was the first recorded person to use the telescope and realised that Copernicus was correct and Ptolemy was wrong. Galileo supported Copernicus's theory of heliocentricism, which stated that the Earth rotated around the Sun, which immediately met with opposition from both a number of astronomers and the Catholic Church. In 1615 Galileo was called to Rome and made to make a public recantation of his 'heretic' claims. He later defended his claims in a paper called *Dialogue concerning the Two Chief World Systems*, which according to the Catholic Church went against the teachings of Pope Urban VIII. Galileo was declared 'vehemently suspect of heresy' and spent the remainder of his life under house arrest, during which time he wrote a number of scientific papers. In 1971 during the Apollo 15 mission, one of Galileo's theories was put to the test by astronaut Dave Scott. He showed that Galileo was right when he said that acceleration is the same for all bodies subject to gravity, when on the Moon and on camera, he dropped a hammer and a feather simultaneously and watched them hit the lunar surface at the same time.

With the invention of the telescope, and subsequently larger and more powerful ones, the discovery of mountains and canyons on these far distant worlds became clearer, and thoughts began of what it would be like to travel to discover what secrets they held. It was at this point that man started to think of how to achieve this and began the first ideas on space travel.

In Germany, the great mathematician Johannes Kepler [1571–1630] went to Denmark to join the Danish astronomer Tycho Brahe [1546–1601] and together they carried out thousands of very accurate positional observations of the planets. When Brahe died in 1601, Kepler used all their findings and published *Kepler's Laws of Planetary Motion* which said:

- A planet moves around the Sun in an ellipse, with the Sun covering one section of the ellipse, whilst the other section is empty.
- The radius vector is equal at all times.
- The square of the sidereal period is proportional to the cube of the planet's mean distance from the Sun.

During the next hundred years or so, numerous fictional stories were written about travelling to the Moon and were in some cases surprisingly quite accurate. In 1657, Francis Godwin, son of the Bishop of Bath and Wells, wrote a science fiction book called the *Man in the Moone*. Godwin was also the Bishop of Hereford and a distinguished scholar, having written a number of theological works. At the same time another Bishop, John Wilkins of Chester, wrote a serious scientific book called *The Discovery of a World in the Moone*, in which he set out to prove that the Moon was inhabited.

As time progressed, more and more scientists and scholars began to take an interest in the stars and in particular the planets closest to Earth. Sir William Herschel astounded the world of astronomy, when in the early 1800s he announced that he had discovered a new planet and named it Uranus, taken from the Greek word *Ourano* meaning 'heaven'. When Sir William died in 1822, his son Sir John Herschel took over his father's mantle, but extended his observations of the stars to the Southern Hemisphere. His father had spent all his life in England and so had only observed the stars in the Northern Hemisphere.

In 1865, Jules Verne published his novel *From the Earth to the Moon*, which described how a capsule, fired by means of a giant cannon, landed on the Moon. This was followed by the *The First Men in the Moon* and both books went into the annals of fictional history.

CHAPTER TWO

History and
Development of the Rocket

The first recorded use of a device that could be used in propulsion was by way of a demonstration in AD 160, by a Greek mathematician and scientist called Hero of Alexandria. The device, called an Aeolipile (Wind Ball), was named after Aeolus the Greek god of wind, and consisted of a rotating hollow sphere filled with water, with two right-angle pipes located 180 degrees apart attached to either end, which were mounted between two supports that carried steam from a closed container suspended over a fire. The jet of steam escaping through the pipes caused the sphere to revolve in the first known demonstration of a crude jet propulsion system. Interestingly enough, there is no record of anyone continuing this experiment until the days of jet propulsion centuries later.

The first recorded use of rockets in warfare was in AD 1232 when the Chinese used 'flying arrows' to repel the Mongols in the battle of Kai-Keng. In AD 1241 the Mongols, quick to learn from their experience, used rockets with devastating effect against the Magyar [Hungarian] forces in the battle of Sejo, which resulted in the capture of Buda – now known as Budapest. In 1258, the Mongols used rockets again when they launched them against Arabs during their capture of Baghdad. [Almost 750 years later rockets were used once again in a battle to capture Baghdad.] Ten years after the Mongols, in 1268, the Arabs used the rocket against Louis IX during the Seventh Crusade.

In Russia, although it is almost certain that rockets were in the country's arsenals, there are no records of their existence until the 1600s, when documents accumulated by the Russian gunsmith Onisin Mikhailov were turned into a compiled document entitled *Code of Military, Artillery and Other Matters Pertaining to the Science of Warfare*. This document contained detailed descriptions of rockets referred to as 'Cannon balls which run and burn'. Although now even the information that existed in this early document has been disputed, mainly because the information which compiled the main manuscript was not entirely Mikhailov's, but was in fact a collection of some 663 snatches of information and articles from a variety of foreign military books and sources. But did that really matter? The fact that Mikhailov collected and published these articles is of more importance than the debate at the time, to try and decide whether or not he was justified in calling them his own work.

Peter the Great of Russia devoted a lifetime to creating Russia's military might, and in 1680 he founded the first Rocket Works in Moscow. There they made illuminating and signal rockets for the army, all of which were under the guidance of English, Scottish, Dutch, German and French officers, who instructed in their use. Then in the early 1700s Peter the Great made St. Petersburg the new capital of Russia, and moved the entire Rocket Works to this new location, expanding it at the same time.

In Poland the commander of the Polish Royal Artillery, Kazimierz Siemienowicz [1600–1651], who was an expert in the fields of artillery and rocketry, wrote a manuscript on rocketry that was partially published before his death. The book *In Artis Magnae Artilleriae pars prima* showed a design for multistage rockets that were to become the basis of technology for rockets heading for outer space. He also proposed creating batteries for launching military rockets and the fitting of delta-wing stabilisers to replace the guiding rods that were currently being used. It was rumoured that Siemienowicz was killed by members of Guilds that were opposed to him publishing their secrets, and they hid or destroyed the remaining parts of his manuscript.

In England, Sir William Congreve, who had been following Russia's development of the rocket very closely, carried out secret experiments at the Royal Laboratory at Woolwich, in developing substantially more powerful rockets. The formation of the Field Rocket Brigade in 1809, under the control of the Horse Artillery, proved their worth during two successive campaigns, Boulogne in 1806 and Copenhagen in 1813, where the Danes were subjected to a barrage of some 25,000 rockets. An explosive shell invented by Lieutenant General Henry Shrapnel supported this 'secret' weapon, which caused devastation to the French cavalry and infantry during the battles against Napoleon. The Rocket Brigade was involved in every campaign around this time, distinguishing itself particularly well in the battle at Waterloo, where Napoleon was finally defeated.

The development of Russian rockets continued into the late 1700s, when an officer in the Tsar's artillery, Alexander D. Zasyadko, who had been studying Congreve's progress and exploits, together with the files from the Rocket Works, decided to design some rockets of his own. So successful were the tests of his rockets that in 1817 Zasyadko was assigned to western Russia to train the Tsar's soldiers in the use of military rockets. The following year a school of artillery was opened and Zasyadko was appointed its head with a promotion to Major General.

In the Russo–Turkish War of 1828–1829, solid fuel rockets were used during the sieges of Varna, Braila, Silistra and Schmia, and Russian warships on the Black Sea used the rockets with great success. During the Crimea War thousands of rockets were used, with increasing devastation and greater reliability. The death of Zasyadko in 1837 caused his position as head of the school to be taken by another artillery officer by the name of Konstantin I. Konstantiniov. This 30-year-old officer was the first to work on the practical problems of rocket production and became the founder of experimental rocket dynamics. Up to this point, the production of rockets was left to the individual skill of the makers, and as can be imagined, there were a number of accidents. Among Konstantinov's achievements were the development of large-scale rocket production and a rocket that could fire lifelines to wrecked vessels.

A number of other Russian inventors produced ideas over the next few years, and then came the innovations of Konstantin Eduardovich Tsiolkovsky [1857–1935], considered by many Russians to be the 'Father of Soviet Space Flight'. The son of a forestry expert and inventor, Tsiolkovsky enjoyed a normal childhood, then when he was 8 years old he contracted scarlet fever and became almost totally deaf. Unable to go to school, he taught himself from his father's books, first mastering mathematics, then physics. After three years at a technical school he returned to his hometown to become a teacher. It was whilst he was a teacher that he started to carry out serious research in the areas of the aeroplane, an all-metal dirigible and a rocket for interplanetary travel. The development of the rocket was the one idea that attracted his attention more than the others and he concentrated his efforts in that direction. However, he still retained an interest in the aeroplane and the dirigible, and produced designs for both.

This was endorsed in 1881, when Nikolai I. Kibalchich [1853–1881] proposed the idea of heavier-than-air machines being propelled by rocket propulsion and carrying human passengers. Unfortunately, Kibalchich also used his expertise in another direction. He was the bomb expert and maker for a revolutionary organisation, and a bomb thrown at Tsar Alexander II, which mortally wounded him, was found to have been made by Kibalchich. He was arrested by the Tsar's secret police and executed soon afterwards.

Tsiolkovsky was a man of vision and was years ahead of his time in his concepts and designs. Between 1885 and 1895, Tsiolkovsky carried out a great deal of work on the design of metal airships and at first he was considered to be an eccentric living in a world of fantasy. In 1894 he proposed a design for an all-metal aeroplane in an article entitled, '*The Aeroplane, A Birdlike Flying Machine.*' Nine years later, in 1903, he

Above left: Konstatin Tsiolkovsky. (Gromov Institute)

Above right: Nikolai Kibalchich. (Gromov Institute)

put forward a paper, 'Investigating Space with Reaction Devices'. He supported his theory by giving practical instructions on the design of rockets and propellants. But he was too far ahead of his time and the paper went unnoticed.

A Russian engineer by the name of Fridrikh Tsander [1887–1933], a devotee of Tsiolkovsky, came to prominence in 1908 when, after corresponding with the scientist, he was asked to edit a compilation of the great man's work.

Over the following years Tsander designed a number of rocket planes, engines and boosters, and developed a reputation for lecturing on the ideas of space travel. So dedicated to the science of space travel, and all that it encompassed, that when he married, he even named his daughter and son Astra and Mercury.

Tsander's expertise covered three areas: the development of the rocket engine, the building and testing of rocket engines and the problems that could be associated with spaceflight. The latter of course covered the problem of escaping from the earth's gravitational field.

The development of rockets to be used in space travel was also an area in which Tsander showed great interest. He favoured using a multi-stage rocket incorporated into an aeroplane. The idea was that the aeroplane would fly the rocket to a high altitude at which point the rocket would then be ignited to push itself into space.

In 1928, Tsander developed the OR.1 (Opytnyi Reaktivny) engine using compressed air and gasoline as the propellants. With the success of this engine, which completed more than 50 tests, work began on developing the OR.2 engine that was powered by benzene and liquid oxygen. The engine was successfully tested in March 1933, and plans were drawn up to put it into an experimental glider known as the R-1. Then tragedy struck. Just ten days after the successful testing of the OR.2 engine, Fridrikh Tsander contracted typhoid fever and died. A memorial to him and his work was erected in the town of Kislovodsk.

His work continued, and his ideas and designs were made available to the science academies, which allowed interested engineers to use them. Amongst these was Yuri Kondratyuk [1897–1942], an engineer from Poltava, Ukraine. Born Oleksandr Ignatyevich Shargei, [he changed his name to Yuri Kondratyuk after the Russian Revolution in 1917] on 21 June 1897, his mother, Ludmila Lvovna Schlippenbach, was said to have been a descendant of Swedish General Wolmar Anton von Schlippenbach, who, in 1709, led the Swedish army's failed invasion of Russia in the time of Charles XII of Sweden. Yuri's father, Ignatiy Benediktovich Shargei, studied mathematics and physics at Kiev University and became a lecturer. Both Yuri Kondratyuk's parents were intellectuals, but his mother, having been arrested on a number of occasions for demonstrating on social reform, was found to be suffering from a mental disorder. Yuri's father divorced her and it was left to his paternal grandmother to bring him up.

On leaving school Oleksandr/Yuri went to The Peter the Great Polytechnic in Petrograd where he studied engineering. At the onset of the First World War he joined the army as a junior officer, but all the time he was fighting at the front his mind was elsewhere. By the end of the war he had filled four notebooks with drawings of spacecraft and calculations on how to send them into space.

The end of the First World War saw the start of the Russian Revolution and a dramatic increase of great poverty in the country. Surviving members of the Tsarist army were considered to be enemies of the state and were hunted down. Oleksandr contracted typhus and was nursed back to health by friends who also found him a new identity – Yuri Vasilievich Kondratyuk. He went to work in Siberia as a mechanic and whilst there, completed his book *The Conquest of Interplanetary Space*. As no publisher would undertake to publish 'this flight of fantasy', as one put it, Yuri published the book himself. Other engineers enthusiastically read it and 2,000 copies were printed.

His engineering expertise came to the fore in the early 1920s, when he designed, and built, a massive grain elevator that was constructed without the use of a single nail. The success of the grain elevator was admired by almost everyone, that was until one of Stalin's purges in the 1930s, when the NKVD [Russian Secret Service, later the KGB and now the FSB] arrested him for being a saboteur. They used the fact that the grain elevator had been built without nails as an excuse to accuse Kondratyuk of planning the structure to collapse. The fact that the structure had been there for some years seemed to have eluded them. Sentenced to three years in a labour camp, his engineering expertise was quickly recognised and he was sent to the Kubass coal-mining region to evaluate foreign machinery. His reputation for the building of the grain elevator had preceded him and he was transferred to Siberia to work on grain projects. It was while here that he became involved with the design of a wind power generator scheme for the Crimea. His design for a 500 ft tower to which was fitted a propeller with a 240 ft span was submitted and accepted. He was sent to Moscow to meet with Sergo Ordzhonikidze, the Commissar for Heavy Industry.

It was on this visit to Moscow that he met Sergei Korolev [1907–1966] the head of the Rocket Research Group [GIRD] who, having read Kondratyuk's book and other works of his, offered him a job. Tempting though it was, Kondratyuk refused, worried that his true identity would be revealed during any security check. Despite this he continued with his research into the launching and recovery of spacecraft. It is surprising that his ideas of landing on the Moon and returning back to earth are very similar to that of the Apollo missions. His theory for re-entry was to approach the earth without reducing speed, and use the atmosphere not only to slow the speed of the spacecraft down, but also to capture the spacecraft using the earth's gravity.

At the outbreak of the Second World War, Kondratyuk joined the Soviet Army and fought against the Germans but disappeared during the fighting near Zasetski. His body was never recovered. Some rumours said that he went to the United States and continued his work under an assumed name, but there is no evidence to support this.

Tsiolkovsky's paper had been re-published in 1911 with additional information regarding propellants, and this time other scientists and engineers took notice. His work was slowly becoming accepted and he was no longer regarded as an eccentric, but it wasn't until 1918 that he was completely recognised for what he had achieved. In

1919 he was elected to the Russian Socialist Academy and granted a personal pension by the Commission for Improvement of the Lot of Scientists [TsEKUBU]. This later became the Academy of Sciences USSR.

A number of groups and societies started to spring up around Russia, their objectives being the investigation into space travel. The government supported them by setting up a department called the Central Bureau for the Study of Problems of Rockets (TsBIRP). The department's role was to bring together all the different societies and groups under one 'umbrella' so to speak, to discover what strides were being made in the west, to assimilate all the results and information and see if they would have any military applications.

In 1924 Tsiolkovsky published a book called 'Cosmic Rocket Trains' in which he described in great detail his design for a two-stage rocket. The first stage would take the rocket through the earth's atmosphere, at which point it would separate and return to earth, whilst the second stage would continue out into space and beyond. He then moved his attention away from rockets to the jet engine, and in 1930 wrote an article explaining the advantages and disadvantages of the jet engine.

Tsiolkovsky died in 1935 a national hero, bequeathing all his papers and models to the Soviet Government. In 1952, the Aero Club of France had a large gold medal struck in his honour, and in 1954 the Soviet government established the Tsiolkovsky Gold Medal, which has been awarded every three years, since its inception, to the most outstanding contributor to space technology.

Two of the main organisations in the late 1920s were the Leningrad GDL [Gas Dynamics Laboratory] and GIRD [Group for the Study of Reaction Propulsion]. The GDL group concentrated their efforts on designing and developing high energy, solid rocket motors. These were developed mainly for military purposes – aircraft rockets and later anti-tank rockets.

The first of the rockets from GDL was the ORM-1 (Opytnyi Reaktivnyi Motor). Designed by Fridrikh Tsander, it was powered by gasoline as the fuel, and liquid oxygen or nitrogen tetroxide as the oxidiser. This was followed by the ORM-2, which was fitted with a solid-propellant ignition and liquid hypergolic propellants. The latter was a propellant that ignited on contact. With the information gathered from these tests, more advanced engines were developed, the ORM-4 to ORM-22. One of the most outstanding engines developed from these was the ORM-52, which was chosen to power a naval torpedo, a meteorological rocket and an anti-aircraft rocket.

Sergei Korolev, who had offered Kondratyuk a job with his rocket research group, found himself on the receiving end of one of Stalin's purges in 1937. He, together with a large number of other aerospace engineers and rocket designers, was arrested and imprisoned. However, with the outbreak of the Second World War, Stalin recognised the need for aerospace engineers and set up a series of prison design bureaus – known as sharashkas. In charge of one of these sharashkas was the aircraft designer Sergei Tupolev and he immediately requested that Korolev be assigned to work with him.

During the Second World War, the Germans created the V-2 rocket (*Vergeltungswaffe-Zwei*: Vengeance Weapon No.2.) which was designed to bring terror to Britain, but fortunately the war ended before it could be brought into full production and unleashed upon Britain. At the end of the Second World War, the strides made by the Germans in the field of rocketry during the war were desperately sought after by the Americans and Russians, and tremendous efforts were made to capture the scientists and engineers involved. The Americans were desperate to get hold of these scientists and engineers and it was they who managed to capture the main people involved – Werner Von Braun and his team. I say capture, but it was the German scientists themselves who sought out the Americans rather than give themselves up to the Russian Army. The Second White Russian Front, commanded by Marshal Konstantin Rokossovsky, entered Peenemünde and started to gather up all what was left of the rockets and facilities, including what documents and drawings the German engineers and scientists had left. Amongst some of the things they found were the blueprints of the A9 and A10 inter-continental rockets, the V-2, and a number of surface-to-surface and air-to-air missiles.

The first Soviet-built rocket, the R-1, was launched on 1 April 1953 and was almost a direct copy of the German V-2, but built with Soviet components. This was supposed to be the first of the Russian ICBMs, but the R-1 only had a range of 3,500 miles. Because this rocket could barely reach the United States, the development of the next rockets, the R-2, -3, -4, -5 and -6, was hurriedly brought forward, as the tension between the two superpowers increased.

Despite this, Russian scientists, together with some of the German scientists and engineers they had captured, started to develop missiles, and from the rockets that launched them, came the first tentative steps to put a man into orbit around the earth.

The minister in charge of armaments, Dmitry Fedorovich Ustinov, was given the responsibility of developing the rocket and missile programme. He had been impressed with Korolev's organisational abilities, as well as his expertise in the field of rockets, and appointed him the head of construction for the development of long-range missiles. With this appointment came the rank of Colonel in the Red Army, but technically Korolev was still a prisoner and subject to the rules governing prisoners.

Korolev ordered the immediate round-up of 200 German workers from the Mittelwerke V-2 factory and had them shipped to Lake Seleger, which was situated between Leningrad and Moscow. In a strange twist of fate, the prisoner now became the jailer and responsible for the security of the facility and its inmates. The Germans had no direct contact with the construction of the rockets or the Russian engineers building them; all their input came from interrogation and written sources.

In the early years it was thought that the development of the Russian space programme was for propaganda purposes and to show the world the strides communism was making. But those who knew Sergei Korolev were well aware of his dedication and the progress he was making in this field. For many years, Korolev's name was kept secret, and only those who needed to know knew that he was the head of the space programme.

Because the war had left the Soviet Union bereft of essential materials usually associated with the manufacture of rockets, Korolev used stainless steel in place of aluminium and titanium, making the construction of his designs heavier than the American ones. This meant that the engines had to be more powerful, and instead of developing large, complicated engines, he developed much simpler engines mounted in several pods around the rocket.

There were also other problems. The engineers selected for the rocket programme came from the military SRF [Strategic Rocket Force] and, unlike the majority of their American counterparts, who in their youth had tinkered with engines and radios, the vast majority of Russians had no experience of this. There was a singular lack of automobiles and radios in the Soviet Union in the early years that prevented young men and women from gaining any form of experience. Therefore the rockets developed had to be of basic design and construction, rugged and reliable to enable them to endure the unsophisticated hands of the engineers and the harsh winters and the extremely hot summers of this vast country.

The Russians up to this point appeared to have been the only ones to show any interest in the serious development of rockets; the British and the Americans appeared to have shown some early interest which then fizzled out, leaving it to a small number of western physicists and scientists to explore this area of science. Among these was the man who was later to be known as the 'Father of American space science', Robert Hutchings Goddard [1882–1945]. In the early years of his experiments he was vilified in the press and regarded as a 'crackpot', an eccentric, and not to be taken seriously. But when the famous aviator Charles Lindbergh became involved, followed by sponsorship from the Smithsonian and later the Guggenheim family, he suddenly became a person of interest and his work quickly became to be taken seriously. It was during his research and projects that he created what was to become a vital piece of military hardware – the Bazooka. With financial support secured, the Goddard family moved to Roswell, New Mexico. It was during this period that he was approached by the German rocket scientist Hermann Oberth [1894–1989], but Goddard refused to cooperate with him and the other German rocket experimenters. Oberth had had Goddard's 1919 paper 'A Method of Reaching Extreme Altitudes' translated into German which was also read by a young Werner von Braun. They realised that Goddard's engine design was capable of achieving efficiencies that were thirty times greater than conventional rockets. The development of the simple Russian rockets had a knock-on effect. The Russians relied for many years on batteries to power their spacecraft, whilst the Americans had progressed to using solar power. This meant that the Russian spacecraft spent much shorter periods in space, which in turn meant that they had to launch considerably more rockets. Because of this they gained vast experience in launching techniques, whilst the Americans, whilst lagging behind in this field, increased the efficiency and reliability of their rockets.

After the war, a large number of German rocket scientists, led by Werner Magnus Maximilian Freiherr von Braun, surrendered to the Americans and were spirited away in 'Operation Paper Clip', also known as the 'Paper Clip Conspiracy', to start and develop a rocket and space programme in White Sands, New Mexico. The remaining scientists left in Germany were taken by the Russians to start building their space programme. The dream of sending a man into space, and ultimately another world, was about to become a reality. When Werner von Braun was taken to White Sands, he took with him the knowledge gained from Goddard's papers. The work carried out by Werner von Braun and his team over the next few years was instrumental in developing America's space programme.

In 1955 America declared their intention to put a satellite into orbit for the International Geophysical Year [IGY], which was scheduled in 1957–1958. The USAF and the CIA had long considered putting a reconnaissance satellite into orbit and took the opportunity to explore the possibility. The National Security Council [NSC] backed the proposal for the IGY satellite as long as it did not interfere with the military programme. In the meantime the US Navy's Research Laboratory [NRL] were involved in a scientific programme of their own using the Vanguard rocket, as opposed to the Atlas and Redstone rockets favoured by the Army and USAF. In September 1955 the Department of Defence [DoD] approved the Vanguard rocket for the IGY programme. The Glenn L. Martin Company was assigned to be the prime contractor as they had also built the Viking rocket, which was a rocket developed for the US Navy for their sounding programme. A number of these rockets carrying scientific instruments were launched to record data about atomic tests. Fourteen of the Viking rockets were built to test the control, propulsion and structure of the rockets, and to test the region of the upper atmosphere that affects long-range radio communication. They also examined the potential of the rocket as a tactical ballistic missile.

Designed as a three-stage rocket, the Vanguard was in essence an uprated Viking rocket. The X-405 liquid-fuelled engine, built by General Electric, was a derivative of the RTV-N-12a Viking engine and powered the first stage. The second stage was powered by an AJ10-37 liquid-fuelled engine, whilst the third stage was powered by a solid-propellant rocket motor. The Vanguard had no stabilising fins and so the first and second stages were steered by gimballed engines. The autopilot, inertial guidance and telemetry systems were housed in the second stage.

The Vanguard rocket had been intended to be the launch vehicle that would put an American satellite into orbit. Initial problems however forced them to use a Juno 1 rocket to place the *Explorer 1* satellite into orbit. From 1957 to 1959 the Vanguard programme used eleven rockets but only three were successful in placing a satellite into orbit. Designated Vanguard TV-0 and TV-1, the first two flights were launched to test the telemetry systems and the separation of the first and second stages. TV-3 was launched from Cape Canaveral, Florida, by the US Navy on 6 December 1957 carrying a 3.3 lb [1.5 kg] satellite. The Vanguard TV-3 lifted off the pad but only reached 3.9 ft [1.2 m] before toppling over and exploding. The satellite was

Robert Goddard with his experimental rocket. (NASA)

blown clear and immediately started transmitting signals. It had been heralded as America's answer to Russia's *Sputnik*, but was a complete disaster and was dubbed 'Kaputnik' by the American press. The Russian premier Nikita Khrushchev, on hearing about the incident, commented about the Americans trying to launch a 'little grapefruit'.

On 17 March 1958 Vanguard 1 [TV-4] was launched, this time successfully, and placed a 3.2 lb [1.47 kg] satellite into orbit. The next four launches were unsuccessful, and it wasn't until 17 February 1959 that the next successful launch, *Vanguard 2*, was made. Two more unsuccessful launches quickly followed, then on 18 September 1958, *Vanguard 3* was successfully launched. This was the last of the Vanguard programme, and although there had been a large number of failures, their contribution had played an important part in the American space programme.

CHAPTER THREE

The Mercury Programme

Mankind's first great step forward into outer space consists of flying beyond the atmosphere and creating a satellite of the earth. The rest is comparatively easy, even escape from the solar system. But of course, I do not have in mind, here, descent on to the giant planets.

Konstantin Tsiolkovsky, 1926.

The 'Space Race', as it became to be known, started on 4 October 1957 with the launch of Russia's *Sputnik 1* [Travel]. The highly polished aluminium alloy sphere had a diameter of 23 in [58.42 cm] and weighed 184 lb [83.6 kg]. It was placed into an orbit with an apogee of 588 miles and an orbit of 96 minutes. One of the pieces of information sent back reported on the physical condition of the upper atmosphere. *Sputnik 1* remained in orbit for ninety-six days transmitting data on the upper atmosphere and the ionosphere. It completed 1,440 orbits of the Earth before burning up on re-entry on 4 January 1958.

The following month, 3 November 1957, saw the launch of *Sputnik 2*, this time with a dog, Laika, as a passenger. Laika was one of three stray mongrel dogs that had been picked up from the streets of Moscow and was the first animal to orbit the Earth. There was no provision for her recovery and she died from overheating and stress just hours into the flight due to a failure of the cooling system. It wasn't until

Sputnik 1 being prepared for launch. (Gromov Institute)

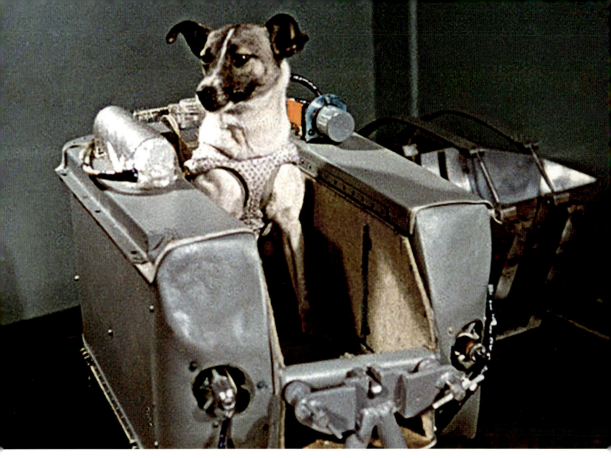

Laika, the first living being to go into space. (Gromov Institute)

2002 that the Russians admitted that was how she died, for up to this point they had always claimed that she had been euthanised with poisoned food. Despite her early demise, Laika provided scientists with the first data on a living organism in a space environment. The first animals to orbit Earth and land safely were on the Soviet *Sputnik 5* mission in 1960, with dogs Strelka and Belka.

The *Sputnik 2* spacecraft carried out 100 orbits of the Earth and transmitted data on the long-term effects of weightlessness on a living animal. It also discovered that as the spacecraft passed over the north and south polar regions of the outer zone of the Van Allen Belt, there was an increase in the flux density of the high-energy charged particles.

When news of this latest mission reached America it was greeted with a sense of shock and disbelief, which reverberated throughout the country, but it had the effect of galvanising the American government into action, and on 31 January 1958, America launched her first satellite, *Explorer 1*.

The satellite was built by the Jet Propulsion Laboratory [JPL] and was launched on top of a Jupiter-C 1 rocket, the brainchild of Werner von Braun and his team of German rocket scientists and engineers. They had been brought from Germany by the US Army Ballistic Missile Agency [ABMA] to work on America's ballistic missile programme at Fort Bliss, just outside of El Paso, Texas. They had been a major part of the Peenemünde team that had developed the V1 flying bombs and

V2 rockets that were launched against Britain during the Second World War, and so were well versed in the development of rockets. *Explorer 1* was the second satellite to carry a mission payload [*Sputnik 2* was the first] and weighed 30.8 lb [13.97 kg] of which 18 lb [8.3 kg] was instrumentation. The satellite was placed into an orbit with an apogee of 2,580 miles [2,550 km] and a perigee of 222 miles [358 km]. The expected lifespan of *Explorer 1* was around three years and it was powered by mercury batteries which powered the high-power transmitter for thirty-one days and the low-power transmitter for 106 days. The batteries finally ran out five months later, but the satellite itself stayed in orbit for the next twelve years and, after completing 58,400 orbits of the Earth, was burnt up as it re-entered the atmosphere on 31 March 1970. There were four more Explorer satellites launched, *Explorers 3* and *4* were successful, but *Explorers 2* and *5* failed to reach orbit and were lost. All these satellites were launched using the Jupiter-C1 rocket, which was later replaced by the Jupiter-C 2.

Rockets were not the only experiments going on around this time, and the North American X-15 experimental aircraft flew right up to the edge of space on a number of occasions. Although there doesn't appear to be much of a connection between aircraft and rockets at this stage, the X-15 played a vital part in America's space programme. During its 199 flights between 1959 and 1968, this air-launched rocket aircraft, which was dropped from a Boeing B-52 bomber, regularly reached heights of 67 miles [108 km] at speeds of 4,520 mph [7,274 kph]. These test flights established important parameters for attitude control in space and re-entry angles. Among the twelve pilots who flew the X-15 was NASA test pilot Neil Armstrong who was later to become the first man to walk on the Moon.

Despite their apparent interest and research into the development of the rocket, their lack of technological experience caused the Russians to launch large numbers of basic rockets, whereas the Americans, with their sophisticated rockets, launched fewer with better results.

At the beginning of October 1958, just one week after the formation of NASA [formerly the NACA], the first administrator, T. Keith Glennan, announced that the first manned space programme called Mercury [the winged messenger of the Gods] was to be created. The principal objectives were to:

- Put a manned spacecraft in orbital flight around the earth.
- Investigate man's ability to function and perform tasks in space.
- Safely recover the man and his spacecraft.

In order to do this a number of guidelines were established, which included using existing off-the-shelf technology, simplified designs, using an existing launch vehicle, a reliable launch escape system and a back-up system that allowed the pilot to take over manual control. Landing the spacecraft would be by parachute and into water, rather than on land, although the Russians always brought their spacecraft down on land. The water landing would increase the safety aspect of the project and, although there were accepted tremendous risks, all the known possible problem areas that could be covered were being explored and addressed. One proposal put forward for landing the spacecraft was by means of a huge inflatable wing, much like today's hang gliders, and the piloted touchdown would be on skids. During trials the para-glider, as it was called, encountered difficulties in getting the wing to inflate fully and had problems in handling during touchdown. The other problem the para-glider faced was the terrain, because despite the best of intentions there was no way a smooth landing area could be guaranteed, so it was decided to continue using the parachute descent into water. To aid recovery of the spacecraft after landing, it was fitted with a UHF recovery beacon, a UHF rescue beacon transceiver and flashing light. The light was designed to be seen from 12,000 feet and from fifty nautical miles on a starless, moonless night. A pulsed UHF output signal was activated on impact, supplying a continuous direction-finding signal on the international distress frequency.

With the relative success of the launch of their Sputniks, the Russians turned their attention to the Moon and on 2 January 1959, using their R-7 heavy lifting rocket, they launched *Luna 1*. The intention was to crash the rocket onto the Lunar surface, but it missed by 3,100 miles [5,000 km]. Undaunted, on 12 September 1959, *Luna 2* was launched and this time the spacecraft crashed onto the Lunar surface on the *Mare Imbrium* basin. Just prior to impact, two round pennants, each with USSR and the launch date engraved in Cyrillic on them, were released from the spacecraft by a small explosive charge, sending them in different directions. This was the first unmanned spacecraft to land on the Moon and a milestone in the history of spaceflight. The R-7 rocket had shown that it was more than capable of launching a spacecraft into

The Paraglider that was put forward as a method of landing the Gemini spacecraft. (NASA)

The Mercury Seven at a press conference. (NASA)

deep space. Then on 4 October 1959 *Luna 3* was launched, only this time it was to orbit the Moon and take photographs of the never before seen far side. Although the photographs it transmitted back were of a very poor quality, they did show that there was a marked difference in the terrain compared with that of the nearside.

In America, applications for the astronaut programme were invited from the military. All applicants had to be under 40 years of age, under 5 feet 11 inches, qualified test pilots with a minimum 1,500 hours of flying time and be in excellent physical and mental condition. Over 500 applications were made which were gradually whittled down to 32, and in February 1959, after extensive tests, the Mercury Seven, as they came to be known, were selected. They were: Captain Leroy Gordon Cooper [USAF]; Lieutenant Malcolm Carpenter [USN]; Lieutenant Colonel John Glenn [USMC]; Captain Virgil 'Gus' Grissom [USAF)]; Lieutenant Commander Walter Schirra [USN]; Lieutenant Commander Alan Shepard [USN)] and Captain Donald 'Deke' Slayton [USAF]. None of the early Russian cosmonauts were test pilots and also had a much lower number of flying hours compared to those of their American counterparts.

At the beginning of 1959 NASA announced that eight Redstone rockets, previously designed to launch missiles, were to be requisitioned from the Army to be used in development flights. At the same time, the McDonnell Company was awarded the contract for building twelve capsules. All the early unmanned rocket tests were carried out from the NASA Wallops Flight Facility on Wallops Island, Virginia. The types of solid-fuel booster rocket used were:

LJ-1 *Little Joe 1*; BJ-1 *Big Joe 1*; LJ-6 *Little Joe 6*; LJ-1A *Little Joe 1A*; LJ-2 *Little Joe 2*; LJ-1B *Little Joe 1B*; BA-1 *Beach Abort*; MA-1 *Mercury-Atlas 1*; LJ-5 *Little Joe 5*; MR-1 *Mercury-Redstone 1*; MR-1A *Mercury-Redstone 1A*; MR-2 *Mercury-Redstone 2*; MA-2 *Mercury-Atlas 2*; LJ-5A *Little Joe 5A*; MR-BD *Mercury-BD*; MA-3 *Mercury-Atlas 3*; LJ-5B *Little Joe 5B*; MA-4 *Mercury-Atlas 4*; MS-1 *Mercury-Scout 1*; MA-5 *Mercury-Atlas 5*.

The Little Joe rocket was used to test the escape system and for re-entry dynamic research, whilst Big Joe was used purely for the development of the heatshield. In September 1959 tests began from Wallops Island, Virginia. The first launch of the Little Joe rocket [LJ-1], carrying a 'boilerplate' capsule and a dummy escape system, was carried out. (A 'boilerplate' was a dummy capsule, the same size and weight as the manned capsule that would be used later, and only used for test purposes.)

The first test of the escape system was a catastrophic disaster, when the system activated prematurely due to an electrical malfunction. The tower blasted off taking the boilerplate spacecraft with it, leaving the rocket still standing on the launch pad. Fortunately the capsule and escape system were successfully recovered, going some way to make up for the initial problem by proving the escape system worked.

Up to this point all rocket tests in America were unmanned, so it was decided to use primates in a series of tests. The use of primates in the world of space exploration was not new: on 11 June 1948, a rhesus macaque named Albert I was placed in a V-2 rocket and launched to over 39 miles [63 km] into the Earth's atmosphere. He did not survive the mission. One year later on 14 June, a second macaque, Albert II, carried out a sub-orbital flight at a height of 85 miles [134 km], but died on landing after a parachute failure. He was the first primate and mammal to actually go into space. The beginning of space, using the Kármán Line, is said to be at 62 miles [100km]. There followed two more flights carrying rhesus macaques, but neither was successful, owing to parachute failures. There followed a number of other tests using animals, with a very small survival rate.

The next primate manned mission was scheduled to be launched on 4 December 1959 using a Little Joe 2 rocket and in its capsule was to be a small rhesus macaque monkey called Sam. The name Sam was an acronym of **S**chool of **A**erospace **M**edicine of the USAF. He was to be placed in a specially created couch with restraints that would help future astronauts handle acceleration forces that would be present during the rocket launch and capsule re-entry into Earth's atmosphere. The protective harness that Sam wore was designed in a way that his hands were able to move, thus showing that future astronauts could work with their equipment and controls during the flight.

On 4 December 1959 *Little Joe 2* [LJ-2] was launched carrying the small rhesus monkey Sam. After reaching an altitude of 55 miles [88 km]and experiencing 19G at Mach 6, the spacecraft flew for eleven minutes before re-entering the Earth's atmosphere, releasing its secondary parachute at 20,000 ft [6,100 m] and its main parachutes

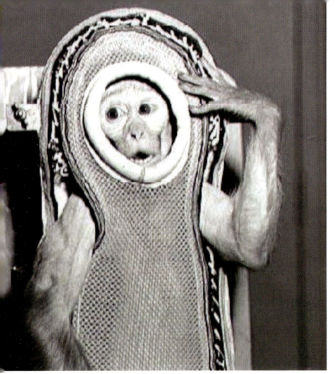

The Rhesus monket Sam in his fibreglass custom-built couch. (NASA)

at 10,000ft [3,050 m]. The destroyer *USS Borie* [DD-704] recovered the spacecraft and little Sam who was found to be in great shape after his space flight.

The use of *Big Joe* [BJ-1] was relatively successful, when, after being launched to investigate the performance of the ablative heatshield, it was recovered safely, albeit 500 nautical miles short of the intended landing site. This was followed by a flight to test its dynamics using *Little Joe 6* [LJ-6], which lasted just five minutes and ten seconds. In November 1959 a second attempt was made to launch an instrumented capsule using an LJ-1A rocket. The main purpose of the flight was to test the escape system under dynamic pressure, but although the launch and subsequent recovery was successful, the testing of the escape system was not. Five more launches using the Little Joe rocket were successful.

The success of these types of rockets prompted NASA to try the Mercury-Atlas (MA) rocket, the first of which was launched from Cape Canaveral on 29 July 1960. The first use of the MA-1 was to test the structure and ablative heatshield during re-entry after the flight had been aborted. But 60 seconds after lift-off structural failure caused the rocket to crash into the sea just east of the launch complex.

The first unmanned flight of the rocket that was to eventually place a man in space, the Mercury-Redstone MR-1, took place on 21 November 1960. Like the first flight of the LJ-1 there was a problem, but one that was to give extremely useful information to the NASA engineers. At the moment of lift-off the launch engine was shut down and the rocket, which had risen no more than a couple of inches off the pad, settled back down. However the escape system had been activated and launched the spacecraft away from the rocket. This was because the Mercury spacecraft atop of the rocket had received an abort signal activating the escape-rocket system, which operated exactly as it had been designed. The cause of the malfunction was discovered to have been caused by two of the ground umbilicals not separating from the launch vehicle in the correct sequence. But at least the engineers now knew that the escape system worked perfectly.

Tests were still taking place on Wallops Island using the Little Joe rockets and again problems regarding the electrical circuits caused concern with the NASA engineers. Little Joe LJ-5 finally lifted off the pad from Wallops Island on 8 November 1960, but during the flight the escape rocket again fired prematurely before the spacecraft had separated from

Ham the chimpanzee in his contoured couch. (NASA)

the launch rocket. The spacecraft remained attached whilst the rocket plummeted into the sea. After a long search the remains of the spacecraft were later recovered from the sea floor, but the condition was such that no evidence as to what went wrong could be discovered.

A second rocket, MR-2, was launched and the recovered spacecraft from the previous launch mated to it. This time the launch was successful and all the tests were completed. This paved the way for the launch of MR-2, this time it was to have a passenger, a chimpanzee by the name of 'Ham' [Holloman Aerospace Medical Center]. The 37 lb [17 kg] primate was fitted into a specially designed biopack seat and installed into the spacecraft atop the MR-2 rocket. The reason behind the selection of a chimpanzee was laid out in a NASA document which said:

> The chimpanzee is a primate that is intelligent and normally docile and of sufficient size and sapience to provide a reasonable facsimile of human behaviour. Its average response time to a given physical stimulus is .7 of a second, compared with man's average .5 of a second. Having the same organ placement and internal suspension as man, plus a long medical research background, the chimpanzee chosen to ride the Redstone and perform a lever-pulling chore throughout the mission should not only test out the life-support systems but prove that levers could be pulled during launch, weightlessness, and reentry.

The flight took place on 31 January 1961, but during the flight the rate of fuel depletion to the launch vehicle was such, that the rocket ran out of fuel before the cut-off velocity system was armed. The result was that the abort signal from the launch rocket was transmitted to the spacecraft. The combination of velocity from the launch rocket and escape rocket motor, caused a far greater escape speed for the spacecraft, which resulted in it landing in the sea over 60 miles [97 km] further downrange from the nearest recovery ship, the destroyer *USS Ellison* [DD-864], than had been planned. A Lockheed P2V Neptune search aircraft located the capsule floating upright in the Atlantic after a three-hour search. But wave action had punched a number of holes in the thin titanium pressure bulkhead of the spacecraft allowing 800lb [363 kg]of sea water to enter through an open relief valve, causing the capsule to slowly capsize. The *USS Donner* [LSD-20], the nearest ship to the landing site, immediately dispatched its helicopter to recover the spacecraft and its occupant Ham. Fortunately the rescuers recovered the spacecraft, with Ham still inside, before it completely filled up.

Ham went to the National Zoo in Washington DC after his brief career as an 'astronaut', where he lived for seventeen years. In 1981 he was moved to a chimpanzee colony in North Carolina where he died in 1983 at the age of twenty-five. He is buried at the International Space Hall of Fame in New Mexico.

There followed a number of other flights using the Little Joe rockets and the majority of these were successful, giving the engineers a massive amount of information to deal with. With the information gathered from these successful launches, the training of the seven Mercury astronauts intensified. Their knowledge of the spacecraft, its design, components and possible problems, had to be almost equal to those of the men who designed and built it. The spacecraft had to be able to protect the astronauts from extreme temperatures, vacuum and the radiation in space. The environment within the spacecraft had to be maintained at a comfortable level even during the high-speed re-entry through the atmosphere.

On 12 April 1961, America was stunned by the news that the Russians had put a man into space. Major Yuri Alekseyevich Gagarin was blasted off from the Baikonur Cosmodrome into space aboard *Vostok 1* and completed one orbit of the Earth before re-entering the Earth's atmosphere. At 23,000 ft. [7 km] Yuri Gagarin ejected from the spacecraft as it approached the landing area, but it wasn't until the 1970s that the Russians admitted that their early cosmonauts ejected from their capsules just prior to landing. Up to this point everyone assumed that the cosmonaut had stayed in the capsule until touchdown. This flight by Gagarin spurred the Americans on to put one of their astronauts into space.

In the United States, a third mission using the MA-3 rocket took place on 25 April 1961, but once again problems dogged the programme when the launch rocket failed to roll and pitch over into the flight azimuth. Flight controllers quickly aborted the mission and were able to save the spacecraft, which suffered only minor damage.

The Mercury spacecraft had been produced by McDonnell Aircraft in their 'clean rooms' and tested in the vacuum plants at their plant in St. Louis, Missouri. Unbelievably there were close to 600 sub-contractors working on the building of the spacecraft, of which NASA ordered twenty units. They were numbered 1 to 20, of which Nos. 10, 12, 15, 17 and 19 were not flown, Nos. 3 and 4 were lost during unmanned test flights and No. 11 was lost when it sank and was recovered thirty-eight years later from the bottom of the Atlantic Ocean. The remaining twelve were either used on the missions or were modified for other experiments.

The unmanned Mercury MA-3's objective was to test the integrity of and the dynamics the rocket would experience during an Atlas launch. The rocket reached a height of 40,000 feet before the parachutes were deployed. The rocket was recovered and the test partially successful, but the test objectives had not been met.

All the Mercury astronauts had taken a keen interest in the development of the spacecraft and insisted that manual control must be included in the design along with windows. There were three ways in which the spacecraft could be controlled:

1. automatically by use of the onboard instruments;
2. from ground stations as the spacecraft passed over using telemetry; and
3. manually.

The latter was to prove to be a necessity, as without it Gordon Cooper's manual re-entry in *Faith 7* on the last Gemini mission would not have been possible. Once in orbit the spacecraft could be manually controlled using rocket propelled thrusters that were fuelled using hydrogen peroxide. This allowed the astronaut to rotate the spacecraft in pitch, roll and yaw positions along its longitudinal axis. To maintain orientation, the astronaut was able to look through the window in front of him, or use a screen that was connected to a periscope fitted with a camera that had a 360 degree rotation.

The single-seat Mercury spacecraft was of a cone-shape design with a cylinder on top. The specifications were: diameter 6.2 ft [1.8 m] across the base, length 6.8 ft [2 m] and on top was an escape tower 9.2 ft [5.8 m] fitted with a solid-rocket motor. In the event of an emergency the escape tower would be fired which would lift the capsule clear of the rocket and parachute it down to safety. The base that carried the heatshield was made of aluminum and covered with multiple layers of fibreglass. Two booster rockets were chosen: the Army Redstone 78,000 lb [35,380 kg] thrust for the suborbital flights and the Air Force Atlas with 360,000 lb [163,293 kg] thrust for the orbital missions.

After launch, if it were not required, the tower would be ejected. The volume of the capsule was just 428.5 cubic feet [13.7 cubic metre]and consisted of an instrument panel with 120 controls: 55 electrical switches, 35 mechanical levers and 30 fuses. In addition the pilot sat in a custom made couch which was slotted in between. To say the capsule was cramped was the understatement of the year.

The astronauts usually flew with their visor on their helmets in the up position, which meant that their suits were not inflated. This made manoeuvrability relatively easy, but with the visor down and the suit, which had its own oxygen supply, inflated, the astronaut could only reach the side and bottom panels, so this is where the vital buttons, switches and handles were placed. The astronaut also had biosensors attached to his chest to monitor and record his heart rhythm, a cuff around his arm to monitor and record his blood pressure and even a rectal thermometer to monitor and record his temperature. All of the data from these attachments were sent to mission control during the flight, where they were constantly monitored and examined. The rectal thermometer was later replaced by an oral one.

There was no onboard computer, so the astronaut relied totally on the computation figures for re-entry and retrofire timings, to be calculated by computers on the ground. Once these had been established and verified, they were transmitted to the astronaut whilst in flight.

Temperatures of over 3,000°F [1,649°C] would be experienced during re-entry through the atmosphere. The ablative heat shield would be burnt off gradually and then, just before landing, the heat shield would be detached automatically, a balloon inflated in its place and parachutes deployed to slow the spacecraft down just prior to impact. Strapped to the convex base was a retropack which contained three rockets that would be activated on reentry to brake the spacecraft speed. Mounted between these rockets were three smaller rockets that were used to separate the spacecraft from the launch rocket at the moment of orbital insertion. At the narrow end of the capsule was the recovery compartment which contained three parachutes: a drogue chute to stabilise the spacecraft's free-fall and two main chutes. The retropack could be severed once the descent parachutes were deployed when it was no longer required.

The Mercury spacecraft's interior had been 'tailor made' for the astronauts with a contoured fibreglass seat, custom moulded to the shape of the astronaut's pressure suit. The pressure suit had its own oxygen supply that supplemented the onboard environmental control system, which also kept him cool. The atmosphere in the capsule was of pure oxygen which avoided the risk of decompression sickness [the bends] and was easier to control. This was to prove fatal later when the oxygen was ignited by a spark causing a fire aboard Apollo 1 and three astronauts died as a result.

Mercury III (Freedom 7)

The scene was now set for the launch of America's first manned space flight – *Mercury III (Freedom 7)*. Initially the mission had been planned for a later date, but the launch of the Russian *Vostok 1* spacecraft on 12 April 1961 had brought the launch date forward. Their planned sub-orbital flight was rapidly brought forward and on 5 May 1961, all was ready. The morning started early for Alan Shepard when he was woken at 1.10am. After breakfasting at 2.40 am he was given a physical examination by a NASA doctor, after which biosensors were placed on his body which had been pre-marked by means of tattoos. He was then helped into his pressure suit by suit technician Joe Schmitt. Schmitt was to suit up all the astronauts for the Mercury, Gemini and Apollo missions.

At 3.55 am, the transfer van, with Shepard on board, left for the launch pad. Whilst inside other suit technicians connected an oxygen supply to his suit in order to purge it. On arriving at the launch site, Joe Schmitt attached the astronaut's gloves to the

Alan Shepard in his spacecraft *Freedom 7*.

pressure suit. Then Alan Shepard, carrying his portable air conditioner, made his way to the elevator that would take him to the Mercury capsule that waited for him on top of the Redstone rocket. Technicians were waiting for him at the top where they helped him into the tiny capsule. Alan Shepard remembered the moment when the hatch was finally closed and he was alone. The biosensors recorded a quickening of his heart rate, but this returned to normal shortly afterwards. He then began the process of de-nitrogenation, by breathing pure oxygen. This was to purge the body of nitrogen, thus prevent an aeroembolism, also known as decompression sickness, a condition sometimes suffered by deep-sea divers – the bends!

The countdown had started, but was put on hold because the cloudy weather conditions were not suitable for photography and it was important that the whole mission was recorded on film. At one point during the launch, when delays were causing frustration amongst the ground controllers, Shepherd's calm voice came over the air, 'Why don't you fix your little problem and light this candle.'

One of the problems, which held up the launch for fifty-two minutes, was to replace a 115-volt, 400-cycle inverter in the electrical system on the launch vehicle. This was followed almost immediately by another problem, when one of the Goddard IBM Computers was found to have an error. It needed a total re-check and configuration to resolve the problem, and coupled with weather delays, Shepard had spent 4 hours and 14 minutes strapped to the couch inside the tiny capsule.

At 09.30 on 5 May 1961, *Mercury-Redstone 3* (MR-3) with Lieutenant Commander Alan B. Shepard, USN, in his spacecraft *Freedom 7*, aboard, was launched from Cape Canaveral. Each astronaut named his own spacecraft but the '7' was added to all subsequent Mercury spacecraft in honour of the first seven astronauts. As the cables and the supporting boom fell away, Shepard started the elapsed-time clock. The lift-off started smoothly enough, but as the rocket passed through the transonic zone, the buffeting became so violent at one point that Shepard's helmeted head was bouncing so hard he was unable to read the dials. Then suddenly the buffeting stopped and the transition to smooth flight passed through without incident. The spacecraft was now reaching a speed of 5,146 miles per hour and Shepard, now in constant voice communication with 'Deke' Slayton in the Mercury Control Centre, informed Slayton that the dial-scanning procedure that had been planned would have to be dropped. Shepard decided that the need to watch his oxygen and hydrogen peroxide indicators was more important than watching the remainder of the dials.

The pressure in the capsule stabilised at 5.5 pounds per square inch as programmed. Just after main engine cut-off, and 2 minutes and 32 seconds after launch, the escape tower jettison rocket fired, followed immediately by the green tower-jettison light on his control panel becoming illuminated.

Outside the spacecraft the temperature had reached 220°F, but the inside capsule temperature was just 91° and his suit temperature was 75°. Shepard then disarmed the retro-rocket jettison switch, then waited whilst the attitude control system rotated the capsule so that the heatshield faced forward. The capsule remained in this position for the remainder of the flight.

Mercury-Redstone MR-3 lifting off the launch pad at Wallops Island. (NASA)

Shepard then began the process of taking over manual control, one axis at a time. First the pitch control axis, which was controlled by means of a hand-controller by his right hand and by moving the controller forward or backwards, Shepard was able to control the spacecraft's up and down pitch. This was then followed by the yaw control, on the same hand-controller, which controlled the left and right motion of the spacecraft by moving the handcontroller left or right. At one point during the flight, Shepard moved his hand to remove a medium grey filter that he had inadvertently left over the periscope, when his hand banged into the abort handle. Very carefully Shepard pulled his hand away, holding his breath whilst he did it. Suffice to say the filter stayed where it was, but the incident did nothing to detract from the breathtaking panorama that spread out beneath him.

As he passed over the peak of his trajectory, Shepard switched to fly-by-wire mode, using the handcontroller to change the capsule's position, aided by the hydrogen peroxide jets of the automatic system. This in effect meant that Shepard would manually place the capsule into position for the retrofire which would take place when the spacecraft reached its zenith at 116.5 miles above the earth.

This was a crucial moment for Shepard, because as he started to make a roll and yaw manoeuvre, he noticed that the pitch position of the spacecraft was between 20 and 25 degrees instead of the required 34 degrees. Concentrating on solving the pitch problem, Shepard suddenly became aware of a gentle 'kick in the pants' as the retrorockets fired and then watched as a couple of pieces of debris flashed by his observation window as his retropack was jettisoned. Glancing at his control panel for confirmation from a green light, he saw nothing and contacted Mission Control. Deke Slayton, as CapCom, said that they had received confirmation of separation, and instructed Shepard to push the manual override. The green light on the control panel suddenly lit up, much to his relief.

As Shepard's *Freedom 7* entered the Earth's atmosphere he was supposed to give Mercury Flight Control altimeter readings between 90,000 and 80,000 feet [27,400 to 24, 400 m], but because his rate of descent was much faster than he had expected he omitted to keep the transmission flowing. At 20,000 feet [6,100 m] the drogue chute snapped out slowing the spacecraft fractionally, then 10,000 feet [3,050 m] later the main chute deployed and the spacecraft slowed reassuringly. Shepard then dumped the remaining hydrogen peroxide fuel and prepared for water impact. When it came it was much softer than he had imagined.

Above the capsule, a US Marine Corps helicopter had watched the spacecraft make contact with the water and immediately moved into position directly above and hooked a cable through the recovery loops. As the pilot raised the capsule to above the water level, another line was dropped and Shepard pulled himself out of *Freedom 7* and into the 'horse collar' on the end of the recovery line. Minutes later the astronaut and his spacecraft were on the deck of the aircraft carrier *USS Lake Champlain* [CV-39]. The flight lasted only fifteen minutes and twenty-eight seconds and reached an altitude of 115.696 miles at a speed of 5,100 mph, but it was the start for which the United States had been waiting. If NASA had listened to von Braun, Alan Shepard would have flown into space on 24 March, beating Yuri Gagarin by three weeks and becoming the first man to go into space.

On returning to Cape Canaveral the following day, both astronaut and spacecraft were examined thoroughly and both found to be in excellent shape. One of Shepard's criticisms about the design of the Mercury spacecraft was that the observation windows were awkwardly placed to observe the stars and because of this he found

President John F. Kennedy addressing Congress with his speech about going to the Moon. (Library of Congress)

himself falling behind in his tasks. Other than this the whole mission went like clockwork and was a complete success. The scene was now set for the second manned flight. Alan Shephard was to be the only American astronaut to fly on Mercury, Gemini and Apollo spacecraft missions.

With the success of the mission still ringing in the ears of the nation, President John F. Kennedy stood before Congress on 25 May 1961, and proposed 'that the US should commit itself to achieving the goal, before this decade is out, of landing a man on the Moon and returning him safely to the Earth.' The space race was well and truly on!

Mercury IV (Liberty Bell 7)

Virgil 'Gus' Grissom had been selected as pilot for the next flight, and on 21 July 1961 his spacecraft MR-4 (*Liberty Bell 7*) lifted off the launch pad. The spacecraft was a modified version of that of Alan Shepard's, inasmuch as it had a modified instrument panel, a larger top window and a side hatch that could be blown using an explosive charge in case of emergency. The pre-launch activities were normal and Grissom was installed in the Mercury capsule to await the launch. Forty-five minutes into the launch countdown, it was discovered that one of the seventy hatch bolts was

Mercury-Redstone MR-4 being readied for launch. (NASA)

John Glenn talking with Gus Grissom just before Grissom entered his spacecraft *Liberty Bell*. (NASA)

misaligned. The launch was put on hold for thirty minutes whilst McDonnell engineers looked at the problem, then decided that sixty-nine bolts would be sufficient to hold and blow the hatch in an emergency. The bolt was not replaced. Alone in the capsule, Grissom experienced some unusual sensations, and one in particular was when the gantry moved back from the Redstone rocket, for a few seconds he felt as if the whole launch vehicle was falling.

At 07.20 am the Redstone rocket lifted off the launch pad and Virgil Grissom said afterwards that it was extremely smooth at first, but severe vibrations started and then seconds later disappeared. At 27,000 feet the cabin pressure sealed off and with relief he watched the suit and cabin temperatures registering normal. The noise of rockets pulling the escape tower clear of the capsule startled him for a moment, but he then watched as the tower drifted off, trailing the remnants of rocket smoke. Then he felt the rocket motor cut off and felt the strange transitional sensation of going from high g to zero g, giving him a tumbling feeling. Ten seconds later a sharp bang indicated that the capsule had separated from the booster vehicle. The capsule started its turnaround process, and although Gus Grissom peered through the window in his spacecraft, he never did catch sight of the launch vehicle. What he did see were the bright rays of the Sun as it streamed through the window, almost blinding him.

All this had taken just under four minutes. It was at this point that Virgil Grissom became a space pilot as he took over manual control of the spacecraft. The view of Earth that was unfolding beneath him was spectacular, so spectacular in fact that he was finding it hard to concentrate on his instruments. He started to carry out the manoeuvring procedures that had been tasked using the handcontroller. He found that the response of the spacecraft to these commands was considerably more sluggish than those of the simulator back on earth.

Gus Grissom continued to be awed by the spectacle beneath him and got behind in his workload. Then suddenly it was time to set the spacecraft for re-entry and at a height of 118.26 miles [190 km] he initiated the retrorockets and seconds later the capsule was arcing downwards. He pitched the spacecraft over into a re-entry attitude of 14 degrees and, as he did so, the glare from the Sun filled his capsule, making it almost impossible to read the dials. This was something that would have to be taken into consideration on future flights, because, had a warning light come on, there is a possibility that the pilot would not have seen it and that could mean the difference between success and failure.

The drogue parachute deployed at 21,000 feet [6,400 m], followed by the main parachute at 12,300 feet [3,750 m]. There was one moment of concern, when Grissom spotted a couple of small tears in the canopy, but they never developed into anything larger. He then felt the landing bag being released ready for impact and then impact, as the spacecraft hit the water. It initially heeled over on to its right hand side, but then gradually straightened up. A heavy swell caused the spacecraft to roll badly but the capsule remained watertight.

Grissom established radio contact with the recovery helicopter's pilot, Lieutenant James, then settled back in his couch and unbuckled his harness. Suddenly the capsule's hatch blew open prematurely and water began to pour in. Grissom realised that there was nothing that he could do and abandoned the space capsule. A heavy swell caused problems for Grissom and he was in danger of drowning and on a number of occasions he went under as his pressure suit began to take on water. He realised that he had not closed his suit inlet valve, but then a second helicopter arrived and dropped a collar to him and he was gratefully winched aboard. The capsule continued to take on water and then, because of the weight of water in the capsule, the recovery helicopter, which by this time had a line attached, was unable to winch the capsule up and so had to let it sink to the bottom of the ocean. Later it was discovered that a warning light in the cockpit of the helicopter had given a false reading indicating a possible engine failure because of the excess strain, with the result that the pilot released the capsule.

A Board of Inquiry was convened and Astronaut Walter 'Wally' Schirra was assigned to carry out tests on the plunger that activated the hatch. After a series of in-depth examinations and practical tests, he came to the conclusion that they would never find out what really happened unless they recovered the capsule. One thing he was sure of, and that was in his opinion, there was only a very remote possibility that the plunger could have been actuated inadvertently by the pilot. He went on to say that during training every astronaut who activated the hatch release, suffered

Gus Grissom being lifted from the sea just after his spacecraft *Liberty Bell* had sunk. (NASA)

a minor hand injury and Gus Grissom did not have any such injury. The Board of Inquiry subsequently cleared Gus Grissom of any misdoings and he was exonerated. The capsule wasn't recovered until 1999 and since then has been on a tour of museums around the United States.

During the debrief, Gus Grissom made a number of suggestions including more egress training in light of what had happened to him. The number of couch restraints should be reduced, the urinal device needed to be more astronaut friendly, the panel lights were far too dim, a point which was highlighted when the sun streamed through the window, and the high frequency communication circuits needed to be replaced as they did not work.

On 6 August 1961, *Vostok 2* with cosmonaut Gherman Titov aboard blasted off from the Baikonur Cosmodrome to send Russia's second cosmonaut to go into space. After completing three orbits of the Earth the spacecraft landed in Krasny Kut, Saratovskaya. At the age of just twenty-six he became the youngest man to go into space, a record that was to last until 2021. Because of the continuing Cold War that still existed between the two super powers, there was some concern by the American people when they realised that Titov's flight took him over the United States three times. Two weeks later NASA's Jet Propulsion Laboratory [JPL] launched *Ranger 1* on an Atlas/Agena B rocket to obtain close-up photographs of the Moon. Known as the Block 1 mission, it was launched on 23 August 1961, but problems arose when the Agena B rocket engine failed to fire. This left the vehicle in a short-lived Earth orbit in which the spacecraft could not collect solar power or stabilise itself and it was subsequently lost. *Ranger 2* was launched three months later on 18 November and that too suffered from the same problem and was lost when it failed to power up and stabilise itself.

In the meantime tests continued with the launch of the fourth flight, *Mercury MA-4*, which was an unmanned flight using a Mercury-Atlas launch vehicle. Launched on 13 September 1961, it was a repeat of the MA-3, but had a number of differences. There were no explosives fitted to the hatch; the large overhead window was replaced with a small viewing window; the instruments were arranged differently and the landing bag had not been installed. The flight was a resounding success and it became the first Mercury unmanned spacecraft to be inserted into orbit and returned to earth. It gave the NASA scientists the information they needed to send another manned flight into orbit. The spacecraft had to have an apogee of approximately 123.3 nautical miles and a perigee of approximately 85.9 nautical miles. During the flight there were a few problems discovered that could affect a manned mission, such as a leak in the oxygen supply system's pressure reducer. Although it was only a small leak and would not have affected the astronaut, steps were taken to immediately remedy the problem.

Despite the success of this mission, it was decided that the next flight, *Mercury MA-5*, would have another animal on board, a chimpanzee called Enos. Launched on 29 November 1961, the mission was for three orbital passes at a perigee of 86.5 nautical miles and an apogee of 128.0 nautical miles. The chimpanzee was given a number of minor tasks, which it completed without any problem. Enos was scheduled to complete three orbits, but the mission was aborted after two due to two issues: capsule overheating and a malfunctioning 'avoidance conditioning' test subjecting the primate to 76 electrical shocks. At the end of the second pass, a problem occurred with the spacecraft's attitude control system, which although being rectified by the ground controllers, caused the spacecraft to use fuel at a high rate and so the third orbital pass was cancelled. After a flight of three hours and twenty-two minutes the spacecraft was brought back to Earth. Two recovery ships were contacted, *USS Stormes* [DD-780] and *USS Compton* [DD-705], both of which had moved into position some thirty miles from the projected landing area. The first to arrive on the scene was

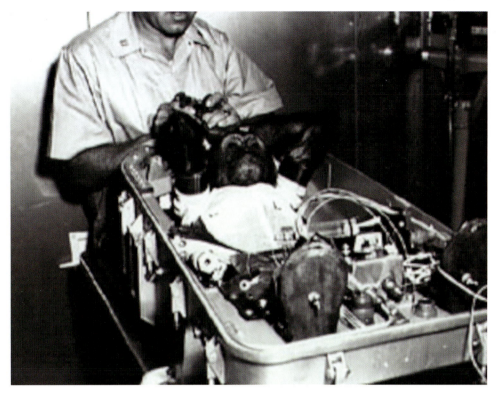

The chimpanzee Enos in his contoured couch. (NASA)

the *USS Stormes* which winched the capsule aboard, cracking the onboard window in the process. On examining the spacecraft it was discovered that the attitude control system had malfunctioned because of contamination in one of the metering orifices. Steps were taken to monitor this on all subsequent flights. Enos was only the third humanoid to orbit the Earth, the first being Yuri Gagarin, the second Gherman Titov.

Meanwhile on 26 January 1962, JPL launched the Block 2 mission *Ranger 3*, and although everything appeared to work well, the spacecraft missed the Moon and was lost because of an inaccurate program. Three months later *Ranger 4* was launched, again using the Atlas/Agena B rocket, but during the flight to the Moon the spacecraft became completely disabled and crashed on to the lunar surface. Despite the loss of the spacecraft, it did prove that the navigation and communication systems worked.

Mercury VI (Friendship 7)

The next flight, scheduled for 20 February 1962, was to be the first American manned orbital flight of the earth. The pilot was to be astronaut John Glenn Jr. As with the previous flight, albeit that was unmanned, the mission was planned for three orbital passes during which time the spacecraft's systems would be evaluated. All the time

John Glenn entering his spacecraft *Friendship 7*. (NASA)

this was going on, the effects of the spaceflight on the astronaut would be monitored by mission control, by means of telemetry.

The early morning start was the same as for the other astronauts, and after a light breakfast and a physical examination, John Glenn suited up with the help of Joe Schmitt. At the Flight Director's position, Christopher Kraft checked to make sure that the world-wide global tracking network was on line and ready.

A hold was called in the countdown because of a fault in the booster's electronic guidance system, but this was quickly rectified. Schmitt assisted Glenn into the capsule and when strapped in, Glenn noticed that a thermistor, which was a respiratory sensor, had moved from its fixed position attached to his microphone. The only way to correct this was for the suit to be opened and this was an extremely tricky procedure to carry out in the capsule. The only other way was to put the launch on hold, for Glenn to exit the spacecraft and be re-suited. It was decided to leave it where it was and ignore the respiratory readings that would be transmitted, as they would be faulty.

As the technicians secured the hatch it was discovered that one of the bolts was broken. A similar problem had arisen in Gus Grissom's flight and it had been ignored. However this time it was decided to replace the broken bolt even though this would place the launch on hold for a further forty minutes. As the repair went ahead the operations technicians took the opportunity to re-check all their systems. With twenty-two minutes to go, another problem occurred when one of the fuel pump outlet valves stuck.

At 9.47 am after two hours of holding, *Friendship 7* lifted off the launch pad and into the history books. The outboard booster engines cut off after just 2 minutes and 14 seconds and dropped away, then twenty seconds later the escape tower was jettisoned. As the G-force increased, John Glenn felt himself being forced farther and farther back into his couch. Then he could hear pieces of loose material hitting the capsule, then watched as they flew past his window. Momentarily he thought that the heatshield was tearing apart, but then realised that it was only a couple of fragments of material. However it did leave a nagging concern in the back of his mind. The jettisoning of the escape tower gave Glenn his first view of the horizon as it dropped away eastward over the Atlantic, and then the posigrade rockets fired and separated the capsule from the booster. Then the automatic control system cut in and turned the spacecraft round to its backward-flying attitude. As the Mercury capsule went into orbit, Glenn noticed the weightlessness as he settled back into his couch with only his restraining straps preventing him from floating. The spacecraft raced along in orbit at a speed of 17,544 miles per hour and Glenn peered through the small window as the world unfolded in a wonderful spectacle below him.

As the spacecraft began its journey around the Earth, tracking and communication stations had been set up to follow the mission. The first station that the capsule passed over was Bermuda, the spacecraft following a north-easterly direction. Bermuda was the only tracking station on the entire network to have radar and

Technician replacng a broken bolt on *Friendship 7* just prior to lift-off. (NASA)

was able to track the spacecraft's insertion into its orbital track, which was vital to ensure the capsule was in the correct orbit. The following were the tracking and communication stations: Maspolomas, Canary Islands; Point Arguello, California; Guaymas, Mexico; Kano, Nigeria; Kauai, Hawaii; Canton Island, Republic of Kiribati; Mercury Control Centre, Cape Canaveral, Florida; White Sands, New Mexico; Muchea, Western Australia; Woomera Test Range, Australia; the ship *USNS Coastal Sentry* [T-ASGM-15] in the Indian Ocean; and in the Atlantic Ocean the ship *USNS Rose Knot* [T-AGM-14]. Of these, five of the stations, Bermuda, Muchea, Kauai, Point Arguello and Guaymas, had a Capcom [Capsule Communicator] and were designated as command posts and were able to track and communicate if necessary with the spacecraft. The reason this track was chosen was because NASA wanted the orbit to pass over friendly countries rather than those who might take exception, believing that they were being spied upon.

The first orbital pass went by with a vivid description of the Earth coming from Glenn, especially as he passed over the Indian Ocean and became the first American to witness a sunset from a height of over 100 miles. As the sun disappeared on the horizon, forty-five minutes later he saw it reappear and commented to Gordon Cooper, who was at the tracking station in Muchea, Australia, 'That was about the shortest day I've ever run into.'

As the spacecraft approached the tracking station in Guaymas, Glenn informed them that he was having a problem with his attitude control system. This was of some concern to Mission Control, because on the flight in which the chimpanzee

USS Coastal Sentry, one of the two tracking ships used during the space programme. (NASA)

USS Rose Knot, one of the two tracking ships used during the space programme. (NASA)

Enos was on board, it had to be aborted because of a similar problem on the second pass. When the spacecraft started to drift at a rate of one and half degrees per second to the right, an automatic signal was passed to the automatic stabilisation and control system thruster, which would then initiate a one pound yaw thrust to push it back. The problem was that this wasn't happening and so Glenn started to switch backwards and forwards between manual and automatic. He found that he was able to correct the drift by switching between the two. After about twenty minutes of doing this, the malfunctioning thruster started working again and everything returned to automatic.

The problem went away for a while, then returned, and despite switching between manual and automatic, it ceased to come back on line. Glenn's test pilot experience then took over, and from then on in, he manually controlled the spacecraft. This caused a number of minor problems because his concentration was now focused on controlling the spacecraft, so many of his observational tasks had to be abandoned. But on the second pass a sensor alarm caused mission control a great deal of concern. The sensor gave an indication that the spacecraft's heatshield had become detached. During the third pass engineers on the ground struggled to find an answer, then decided to keep the expended retropackage, which normally would have been released just before re-entry, attached in an effort to hold the heatshield in place. All tracking sites were ordered to monitor the instruments that were associated with the heatshield, but in their conversations with John Glenn, they were not to mention it.

It wasn't long before Glenn realised that something was up, as every time he passed over a tracking station and became in voice contact, he was asked the same question regarding the switch that controlled the deployment of the landing-bag and was it in the 'off' position.

It was during the second orbital pass that John Glenn observed the 'fireflies' outside his spacecraft and wondered if they had anything to do with the gas being exuded form the reaction control jets. He later determined that they had nothing to do with the reaction control system [RCS].

On his last orbit, Glenn was told to place the landing bag deploy switch in the automatic position and if the light came on he was to complete re-entry with the retropack in place. On hearing this, John Glenn immediately realised the reasoning behind the questions about the switch. He complied with the request and carried out the test. He reported that no light came on and that he had heard no unusual sounds coming from that quarter. It was decided to keep the retropack in place throughout the re-entry until the g reading was showing 1.5.

The spacecraft re-entered the earth's atmosphere whilst everyone held their breath, and because of the three-minute communication blackout that always occurs during this period, this was a further tense moment. During the hottest point of re-entry, John Glenn saw chunks of material flying past his window and admitted later, that the nagging thought he had just after launch when he saw the pieces of material fly past his observation window, was possibly right and that his heatshield was disintegrating. But as the seconds passed he realised that the heatshield was holding and then as he reached 28,000 feet [8,530 m], the drogue chute suddenly deployed. Then at 17,000 feet [5,180 m] the main chute came out and Glenn operated the impact bag switch, and watched with relief as a green light suddenly came on confirming its deployment.

The moment of splashdown was of great relief and the destroyer *USS Noa* [DD-841], which had been monitoring his descent visually, raced to pick him up. A whaleboat was launched, and once alongside the bobbing spacecraft, secured it until it could be lifted on to the destroyer's deck with astronaut John Glenn still inside. The mission had been a complete success.

After a brief medical examination John Glenn was transferred by helicopter to the main recovery ship the aircraft carrier *USS Randolph* [CV-15]. The spacecraft was returned to Cape Canaveral where it was thoroughly examined by engineers looking for the cause of the retropack sensor alarm. It was discovered to have been a fault in the wiring, which resulted in a complete rewiring design to prevent the problem re-occurring.

One of the first stages in the proposed Apollo programme took place on 27 October 1961 when a *Saturn SA-1* first-stage booster with a thrust of 1.3 million pounds [590,000 kg] was launched successfully from Complex 34, Cape Canaveral. Weighing 925,000 pounds [420,000 kg], the rocket carried a water-filled dummy upper stages to a height of 84.4 miles [136 km]. But it wasn't until 25 April 1962 that the next test took place when *Saturn SA-2* was launched. The SA-2 carried 22,800 US gallons in its upper stage, which was released into the atmosphere to investigate the effect it would have on radio transmissions and changes in the local weather. The mission was repeated on 16 November when *Saturn SA-3* was launched.

In the meantime *Mercury 7 (Aurora 7)*, the fourth Mercury Atlas flight, MA-7, was being prepared for launching. This could have ended in disaster had it not been for the stringent medical examination to which the astronauts were subjected. The astronaut selected for the flight was Captain Donald K. 'Deke' Slayton, USAF, but it was discovered during a routine medical that he had a slight heart murmur. Such were the stringent requirements regarding the health and the physical condition of astronauts that his place was taken immediately by Commander Scott Carpenter, USN. Such was the secrecy surrounding the Mercury programme, that Scott Carpenter, after having applied to join the programme, received a letter whilst at sea, from the Bureau of Personnel in Washington, that simply said:

> You will soon receive orders to OP-O5 in Washington in connection with a special project. Please do not discuss the matter with anyone or speculate on the purpose of the orders, as any prior identification of yourself with the project might prejudice that project.

On arriving in Washington he was informed that he had been selected for astronaut training and to report to Cape Canaveral. Then the training began and on completion Scott Carpenter waited for the moment when he would be selected to fly a mission and that came unexpectedly. Scott Carpenter had been the back-up pilot for John Glenn, so he had been in training for Glenn's mission, but Deke Slayton's mission was marginally different. The training became very intense as the 'window' for the launch was much shorter than normal.

Despite three fifteen-minute holds due to ground fog, the launch on 24 May 1962 was the smoothest one to date and was watched on television by an estimated 40 million people. At 38,000 feet [11,580 m] the escape tower jettisoned and Scott Carpenter watched it cartwheel as if in slow motion, away towards the horizon, with smoke still trailing from its three rocket nozzles. Scott Carpenter commented to Gus Grissom, who was Capsule Communicator [CapCom], that he had felt the lift-off, to which he received a terse reply of 'Roger that'! The Mercury Control Centre [MCC] and the outlying monitoring stations were constantly being monitored by flight director Christopher Kraft, who was well aware that this was only the fourth manned mission in the American space programme.

Then there were two bangs as the explosive bolts fired to separate the spacecraft from the booster and the posigrade rockets fired to propel it clear. Scott Carpenter then decided to turn the spacecraft round to its backward flying position manually. John Glenn's spacecraft had been turned automatically and had burned 5 pounds of fuel in doing so. Carpenter used only 1.6 pounds of fuel by doing it manually.

With the spacecraft now in orbit, Carpenter, like Glenn, commented on the fact that he felt no sensation of speed, despite the fact that he was now travelling at 17,549 miles per hour [28,242 km per hour].

Scott Carpenter about to enter his spacecraft *MA-9 Aurora 7*. (NASA)

It is interesting to note that the food take by Scott Carpenter on his flight differed greatly to that taken by John Glenn. Glenn had taken squeeze tubes of the baby-food type, whereas Carpenter had had a variety of foods prepared by the Pillsbury Company that consisted of chocolate, figs and dates. The Nestlé Company had made some bon-bons composed of high protein cereals, orange peel with almonds and cereals with raisins. All these foods were covered with an edible glaze. It soon became obvious that in a weightless environment the squeeze tube variety was the most practical, as when Carpenter bit into one of the bon-bons, the crumbs floated around inside the capsule. This may sound innocuous, but it could have been potentially lethal if any of the crumbs had got behind the electrical panels and caused a short circuit. He also found it difficult to eat with a gloved hand whereas the tube just required squeezing.

Settling down to the flight, Scott Carpenter reported to the various tracking stations as he passed overhead giving them the readings from his instruments, whilst they monitored the telemetry from the spacecraft and from Carpenter himself. He reported having great difficulty in loading the special film into the camera because of the gloves. However he managed to do it and got some superb shots of the horizon whilst over Nigeria.

Over the Indian Ocean, Scott Carpenter carried out a series of visual tests of the stars, but found it almost impossible because of light seepage coming from the instruments in the capsule. As he passed over the Great Victoria Desert near Woomera, Australia, an experiment of firing four flares, each of one million candlepower and burning for 1½ minutes, failed because of cloud cover. It was hoped that the light could have been seen from space. Two passes over the site revealed nothing, so the project was abandoned.

Problems started to arise in the flight, partly due to events outside of Scott Carpenter's control, and others due to his over-eagerness. Taking over manual control of the spacecraft, Carpenter rolled and yawed the capsule in every way he could, including at one point standing it on the antenna canister end. This was in addition to carrying out the designated experiments that were part of his mission brief. Then he inadvertently activated the sensitive-to-the-touch, high-thrust attitude control jets. This caused both the automatic and manual systems to become temporarily redundant. Carpenter appeared to be oblivious to the fact that he was using up fuel at an alarming rate, and flight control were having to continually remind him of this.

The last orbit of his flight was possibly the most relaxing, as Carpenter had to conserve fuel because the tanks were only half-full and so the spacecraft was allowed to drift. He continued to take photographs of the sunset and sunrise and carry out other unauthorised activities, but as he approached Hawaii he was instructed to start his pre-retrofire countdown and shift control from the manual to the automatic stabilisation and control system. Carpenter realised that he was travelling at five miles per second and was way behind on his checklist.

Halfway through his third orbit, Carpenter used his camera to catch the phenomenon of the 'flattened' Sun at sunset. He took almost twenty photographs of this spectacular sight and his description of the sunset endorsed this.

The sunsets are most spectacular. The Earth is black after the sun has set. The first band close to the earth is red, the next is yellow, the next is blue, the next is green, and the next is sort of a – sort of a purple. It's almost like a brilliant rainbow. These layers extend from at least 90 degrees either side of the sun at sunset. This bright horizon band extended at least 90 degrees north and south of the position at sunset.

On the third orbital pass, Carpenter accidentally bumped his hand against the hatch and suddenly a cloud of particles raced past his window. He suddenly remembered John Glenn commenting about seeing 'fireflies' outside his spacecraft, so he tapped the hatch again and another shower raced past. It became obvious that because of the freezing temperature outside the spacecraft, there was a frost coating it and Glenn's 'fireflies' suddenly became Carpenter's 'frostflies'.

As he approached Hawaii, the ground control there instructed him to pass control of the spacecraft from manual to the automatic stabilisation and control system. Aligning the spacecraft ready for re-entry, Carpenter suddenly found that the automatic stabilisation system would not hold the 34-degree pitch and zero degree yaw attitude required. Falling farther behind in his checklist, he switched to fly-by-wire control, but forgot to switch off the manual system. For the next ten minutes the spacecraft used up fuel from both systems unnecessarily.

When the time came for him to hit the retrofire button, Carpenter thought that he had managed to align the spacecraft satisfactorily, but the spacecraft's attitude was about 25 degrees to the right and, as he was to discover later, this would cause an overshoot of about 175 miles [282 km]. This, coupled with the fact that the retrorockets began firing three seconds late, and that the thrust from the rockets themselves was not as powerful as it should have been, an additional 60 miles [97 km]was added to the overshoot.

As the rockets fired, Carpenter realised that he was still on manual control, but the fuel gauges were now reading empty on manual and only 15 per cent on automatic. Then he was in to re-entry and everything became academic. The communications blackspot suddenly came and all transmissions ceased, Carpenter felt very alone, then the spacecraft started to oscillate wildly. At 25,000 feet [7,620 m] he released the drogue chute, and as it deployed the spacecraft slowed fractionally. The main chute was armed at 15,000 feet [4,600 m] and deployed manually at 9,000 feet [2,740 m]. Then came the comforting sound and reassuring jolt of the spacecraft hitting the water – 125 miles [201 km] off the planned landing target!

The spacecraft lay in the water at an angle and Carpenter decided to wait to see if it would stabilise. After a few minutes he realised that this was not going to happen and that the spacecraft was starting to lie quite deep into the water. Alan Shepard, who had been CapCom [Capsule Communicator] throughout the mission, had told him just prior to landing in the water, that because of his overshoot, it would be at least an hour before an aircraft could get to him. Carpenter decided that it was too risky to blow the hatch and so decided to exit the spacecraft though the neck. After a great deal of difficulty, Carpenter struggled through the top of the spacecraft and inflated a yellow life raft. Clambering into it, he waited for the rescuers to arrive, well aware that the raft carried no radio. Fortunately, although there was a heavy swell, the weather was not inclement. Afterwards almost an hour passed before the destroyer USS Farragut [DD-348] arrived and hove-to close by. Then, after about half-an-hour, the drone of an aircraft's engines filled the skies and a P2V appeared followed closely by, of all things, a Piper Apache. It soon became obvious that the Apache was just there to take photographs. Then after twenty minutes an SC-54 aircraft arrived and out dropped two frogmen, who quickly swam to Carpenter in his life raft. They quickly inflated two other life rafts that they had brought with them and locked the three together.

An Air Force SA-16 seaplane had been launched from Roosevelt Roads to pick up the astronaut, but as they approached the scene, the pilot decided not to put the aircraft down because of the choppy seas. It was decided to carry out the pick-up using a helicopter and a ship as had been originally planned.

After over three hours of sitting in his life raft, Carpenter, and the two frogmen, were picked up by an HSS-2 helicopter from the aircraft carrier USS Intrepid [CV-11] and taken back to the ship, none the worse for their ordeal.

The spacecraft itself was picked up by the destroyer *USS John R. Pierce* [DD-753] and returned to Cape Canaveral for its post-flight examination. When it was opened it was found to have 65 gallons of sea water inside. Carpenter remembered some small amounts of water being shipped in, but nowhere near the amount found. A detailed examination of the spacecraft found it to be in good condition despite the usual dents and scratches on the external casing.

There were recriminations about the delay in picking up Scott Carpenter, but most of them were levelled at the astronaut himself. The overshoot was down to him and he was the first to admit this. But to be fair to Scott Carpenter, he had had less than ten weeks to prepare for his flight and this was a major contribution to his problems. Fortunately he and the spacecraft were recovered successfully, but because of the criticism levelled against him by the NASA hierarchy, he was never selected to fly in space again. Instead he turned his talents into becoming an aquanaut in the US Navy's 'Man of the Sea' programme.

Mercury VIII (Sigma 7)

Despite the problems that beset MA-7, the mission itself was a success and so the next mission, MA-8, was scheduled to carry out six orbital passes. The pilot selected for this mission was Walter M. 'Wally' Schirra, Jr. On 3 October 1962, Schirra slipped into his contoured seat in the *Mercury 8, MA-8*, call sign *Sigma 7* and started to carry out all his pre-flight checks. Whilst in the process of stowing away all his bits and pieces ready for the flight, star charts, camera, etc., he found a steak sandwich placed in one of the compartments. Ruefully smiling to himself and thinking of the tubes of tasteless food that had been assigned to the programme, he thought at least he would have something substantial to eat whilst in space.

As with all the other Mercury flights, the countdown started well enough, but then came the 'hold' signal. One of the radar sets at the tracking station in the Canary Islands had malfunctioned and it was a necessary part of the mission that this particular one was up and running. Fortunately it only took 15 minutes to fix the problem and the countdown was restarted. At 07.15 hours the pad rumbled under the power of the Atlas LV-3B rocket and then *Mercury 8* was climbing slowly upwards and heading into space. Then 10 seconds later the rocket developed an unexpected clockwise roll. Seconds later the roll stopped, much to the relief of Wally Schirra and everybody in the flight control, as the launch was only seconds short of reaching abort condition. With the rocket now stabilised it continued on upwards.

Schirra, well aware of what was happening, had his hand on the abort handle and had the roll not sorted itself out, he would have pulled it. The booster engine having cut in pushed the spacecraft onward and upward, then the booster engine cut off and seconds later the escape tower separated as its rockets ignited. The sustainer engine continued and then shut down and with that *Sigma 7* settled into orbit at 176 miles [283 km] above the earth and at a speed of 17,557 miles per hour [28,255 km per hour].

With his eyes glued to the instrument panel, Schirra transferred control to the fly-by-wire mode and started to obtain his correct orbital attitude position. Realising the problems that had risen regarding the excessive use of fuel during Scott Carpenter's flight, Schirra ensured that all his control movements of the spacecraft were slow and deliberate. With the spacecraft now in a stabilised orbit, Schirra allowed himself the luxury of a few minutes of sightseeing as he passed over the African continent.

It was whilst passing over Africa that Schirra suddenly started to feel warm, and back at Mercury Flight Control, they too had been noticing his suit temperature climbing. In fact they were considering terminating the mission and returning Schirra to Earth after noticing that the suit temperature had risen to 82 degrees. The flight surgeon decided that they should give it one more orbit before making a final decision. Schirra slowly adjusted the control valve on his suit, which was at position 4 until it reached position 7, and slowly but surely, the temperature of the suit equalised with his body temperature.

Schirra placed control of the spacecraft on automatic and started to carry out some of the experiments that had been built into the mission. As the night sky rushed to meet him, he aligned the periscope toward the night sky, but like the previous astronauts found the periscope a 'useless piece of baggage'. One of his tasks was to carry out a series of alignments when the spacecraft yawed from the flight path. He also carried altitude correction from a visual aspect, using the reticule in the window. During the night pass over Australia on his second orbit, Schirra carried out one of the hardest experiments of the flight. By using celestial navigation in conjunction with the star-finder charts, he was to carry out an alignment of the spacecraft by positioning it in relation to known stars or planets, then by watching the apparent motion of the stars he would test his sense of facing left or right.

The experiment worked successfully, but was extremely time consuming and tiring. Nevertheless it was an experiment that was very necessary, as by its very nature it would be an extremely important skill to have in the event of an emergency.

When it came to making a yaw manoeuvre under manual control, Schirra used the Moon as a reference, which worked perfectly despite the limited view from his cabin window. Switching the spacecraft back to manual, he noticed that his suit temperature had dropped, making things more comfortable for him.

During his flight over Hawaii, Schirra put the spacecraft into automatic or 'chimp configuration' as he called it and had a chat with Gus Grissom and the Hawaiian tracking station. As Hawaii faded and the Californian tracking station came on line, Schirra noticed the 'fireflies' that John Glenn had reported seeing on his flight. Glenn, who was the CapCom in California, was delighted, as no one other than himself and Gordon Cooper had ever seen these strange phenomena. On his third orbit, Schirra retrieved his camera and started taking photographs of whatever took his fancy, coupled with another series of experiments. On the fourth pass, he went live on radio and TV for a few minutes. The fifth pass was the one that let Schirra relax and take things easy. All the experiments had gone well and the transition from automatic control to manual control and back had gone smoothly, in fact things were becoming rather mundane, but Schirra knew that on the sixth orbit he was going to have to make preparations for re-entry. Schirra noted that the one thing that became very apparent was that there was a rapid 'speeding of time' as the spacecraft raced around the earth.

One of the experiments he had been asked to carry out was a psychomotor experiment, which had been designed to test an astronaut's orientation whilst in space. Schirra selected three dials, then closed his eyes and attempted to touch them. Out of nine attempts he only got it wrong three times, and the worst one was only two inches off target. His conclusion was that an astronaut's orientation would not be adversely affected by weightlessness if having to reach blindly for the controls.

As he approached Zanzibar, Schirra was asked to keep an eye out for the giant Echo balloon-shaped satellite, but he never saw it. He did at one point exclaim that he was flying 'upside down' as he approached the Eastern side of the United States and took a number of photographs. (There is no up or down in space.)

Some of the conversations between the CapCom were broadcast live on radio and television, so both parties had to be careful about what they said, but when in private the conversations were more relaxed. They even joked about using the bungee cord exerciser, commenting that it wasn't exactly like walking around. Deke Slayton's reply was 'Did you say you'd like to get up and walk around?' Schirra's reply is not on record!

Approaching his final orbit, Schirra started to stow everything that was not required and go through his pre-entry checklist. Everything had gone so smooth, that he still had time to carry out one more orientation test, this time selecting the emergency handle as his target.

As he passed over the Indian Ocean tracking station ship *USNS Coastal Sentry* [T-AGM-15] he was asked if he required any help completing the pre-retro sequence checklist, to which he replied 'Negative'. Approaching the Pacific Ocean, he made contact with the tracking station ship *USNS Rose Knot*, with Alan Shepard on board, and reported all was well and that everything was on automatic and he had the manual-proportional system on standby. Alan Shepard started the countdown for the firing of the first retrorockets. Taking the spacecraft off automatic mode and shifting it to fly-by-wire, Schirra prepared the spacecraft for retrofire. At the right moment he fired the retrorockets and the spacecraft slowed markedly. As the spacecraft entered the atmosphere, Schirra watched the altimeter and as it reached the 40,000 feet [12,190 m] mark, punched the drogue chute button. As it deployed, he activated the fuel jettison switch to get rid of the excess fuel. At the 15,000 feet [4,570 m] mark he manually deployed the main chute and watched with great relief as it streamed out, unfurled and then snapped open at 10,500 feet [3,200 m]. The spacecraft gently impacted with the water 4.5 miles [7.24 km] downrange from the intended impact point, just as swimmers and helicopters from the aircraft carrier *USS Kearsarge* [CV-33] arrived.

The swimmers placed a flotation collar around the heatshield and prepared to get Schirra out of the spacecraft. However, feeling relaxed and comfortable, Schirra said that he would prefer to stay inside and have a small boat tow him to the carrier.

Wally Schirra being assisted from his spacecraft after his spaceflight. (NASA)

Within half-an-hour the spacecraft, with Schirra still inside, was lifted onto the deck of the aircraft carrier.

During the medical examination on the aircraft carrier, it was discovered that although showing no ill effects from his space flight, there were some unusual symptoms. It was noticed that when his blood pressure was taken whilst lying down, his heartbeat averaged 70 beats per minute, but taken when standing up, it rose to over 100. It was also noticed that his legs and feet were a reddish-purple colour, which indicated that his veins had become engorged. The following morning his blood pressure had returned to normal and the skin on his legs and feet had resumed their normal colour. The problem with his suit temperature was discovered to have been caused by drying out of the silicone lubricant on a needle valve that controlled the flow of the suit coolant. The flight was textbook and the information gathered from the experiments worth their weight in gold.

Back at JPL, the third of the Block 2 missions, *Ranger 5*, was launched on 18 October and this too was a failure as it missed the Moon completely by 750 kilometres and went into orbit around the Sun. It was discovered later that one of the gold plated diodes used in the five missions was found to be prone to flaking whilst subjected to the conditions of being in space. It was thought that this was what might have

MANNED AND UNMANNED FLIGHTS TO THE MOON

Mercury- Atlas MA-8 spacecraft about to be recovered by naval crew. (US Navy)

caused the problems, because once it was replaced with a different type for the subsequent missions, there were no more problems. The tests of the Saturn-Apollo rockets continued on 28 March 1963 with the preparation of the launch of *SA-4*, but was aborted after a premature shut down of a single S-1 engine. This was also the last time Complex 34 was used and the remaining launches were from Complex 37B.

The last of the Mercury flights 'Mercury IX (*Faith 7*)' was on the 15 May 1963 and was flown by astronaut Gordon Cooper. The launch went as planned without the usual expected delays and at 08.04 EDT, *Faith 7* lifted off the pad and climbed skywards. Two minutes and 14 seconds later, Gordon Cooper heard a sharp thud as the booster engines cut off, the escape tower blasting clear of the capsule following.

The sustainer engines continued to push the rocket toward the atmosphere, the guidance system controlling the flight for the two minutes before the Sustainer Engine Cut Off (SECO) sequence was activated. Just three minutes after launch and with the cabin pressure sealed, Gordon Cooper called Mission Control with the words: '*Faith 7* is all go'.

Once in orbit Gordon Cooper switched control from automatic to fly-by-wire and hurtled over Bermuda at 17,547 mph [28,239 kph]. Before he knew it tracking stations were calling him one after another whilst he worked frantically to confirm the

Gordon Cooper getting in to his spacecraft *Mercury-Atlas MA-8 Faith 7*. (NASA)

telemetry information requested. The speed at which he was travelling was something he was going to have difficulty in coming to terms with, as all the previous astronauts told him. Slowly but surely he became accustomed to carrying out his tasks while his spacecraft raced around the earth.

On the second orbit, Mercury Flight Control informed him that his orbital parameters looked so good, and that all the telemetry signals from his spacecraft were functioning so well, that they had scheduled his flight for a possible 20 orbits. Cooper discovered that his suit temperature was fluctuating erratically, but not enough to cause him any major concern. For a while he forgot everything as he marvelled at the unblinking stars in space and the different weather systems moving around the Earth. As he passed over Guaymas, Mexico, Gus Grissom, who was the capsule communicator for that area, passed the word that he was officially go for seven orbits.

When the third orbit came around, Cooper started to check on some of the experiments for which he had been scheduled, eleven in all. One was to eject a six-inch diameter sphere fitted with polar exon strobe lights. The idea was that on his next pass he would be able to test his ability to spot and track this flashing beacon whilst in orbit. After launching the beacon he scanned space on his fourth orbit, but he could not see anything remotely like a flashing beacon, but on the fifth pass there it was! With unreserved delight, Cooper reported, 'I was with the little rascal all night.'

Cooper continued to keep check on all his medical requirements, ie. blood pressure, temperature, urinary samples and even calibrated exercise. As he passed through his seventh orbit, Cooper was in the process of transferring the urinary samples from one tank to another, when he was made aware that he had passed Wally Schirra's record of six orbits. He was also having problems pumping the urine samples from one tank to the other. He placed on record: 'The thing about this pumping under zero g is not good. (Liquids) tends to stand in the pipes and you have to actually forcibly force it through.' He was also still having problems with his suit temperature, which was becoming more erratic, but then it seemed to level off and caused him no more immediate problems.

The next communication was as he passed over the Zanzibar tracking station, when he was told to go for a further ten orbits, making seventeen in all. So on track was the spacecraft and the actual orbit parameters were so close to those planned, that the measurement between them was down to less than a tenth of a mile and hundredths of a degree. In fact most of the tracking stations and telemetry command ships were becoming very complacent. At one point, after being in space for over thirteen hours, he flew over the area under the control of John Glenn and was told by Glenn to 'Tell everyone to go away and leave you alone'. In short he was told to get some sleep. Cooper told him that he was too excited to be sleepy as there was so much to see and do, but Glenn insisted and Cooper realised that he had to have some rest. Over the next ten hours Cooper napped intermittently, and in between took more and more photographs.

The flights continued as planned and by orbit 15, the only problem Cooper was having was that the heat-exchanger in his suit kept fluctuating between hot and cold. Orbits sixteen, seventeen and eighteen passed without incident, just more experiments, checking and photographs, a number of which were infrared weather ones. He continued with his Geiger counter measurements for radiation and aeromedical experiments throughout, but was finding it increasingly difficult to keep up with all the chores. Amongst the experiments he was tasked with was to look out and photograph, any unusual activity on or around Cuba (this was just after the Cuban Missile Crisis) and look for possible missile sites using a special long range camera built into the spacecraft. At one point Gordon Cooper commented to mission control that all he appeared to be doing was taking pictures, pictures and more pictures and quoted that he was up to 5,242. It was whilst going over the Caribbean looking for large metal objects that shouldn't be there, that he started to notice objects in the shallow waters that he concluded could only be shipwrecks possibly from the days of the Spanish colonial period when they sailed their ships through this area on their way back to Spain. He charted the positions of these possible wreck sites every time he passed over them. Gordon Cooper never pursued his treasure hunt, but his friend Darell Miklos is in the process of trying to find out if there was any treasure to be found.

Cooper's relaxed ride through space suddenly turned sour, when, during the nineteenth orbit, a small green light appeared on the instrument panel, indicating that the spacecraft was decelerating. Cooper reasoned that this had to be a false indicator, as he felt, together with his loose equipment, weightless. On the ground tracking station over California, they had had no such indication, but in the Mercury Control Center there was concern. If there was a problem, it could affect the attitude stabilisation at the point of retrofire.

The problem was further aggravated, when, on the next pass, all the attitude readings were lost. Mercury Flight Center were showing serious concern by this time, but this was nothing to the concern they showed on the next pass, the twenty-first orbit, when a short circuit occurred in a busbar that served the 250-volt inverter. This left the whole of the stabilisation and control system without electric power. There really was now a very, very serious problem.

The radio waves suddenly became alive with transmissions between the spacecraft, the Mercury Control Center and the tracking stations, as *Faith 7* passed overhead. It was also noted that the carbon dioxide level in both the suit and the spacecraft was rising noticeably, highlighted by Gordon Cooper's remark passed to Scott Carpenter in Zanzibar, 'Things are beginning to stack up a little'. This really was an understatement. Cooper then said that he intended to make a manual re-entry.

The time had come to bring the spacecraft home. Gordon Cooper remained remarkably calm throughout the problem and by the time he had completed his checklist with John Glenn, all was ready to bring the spacecraft back under manual control. The re-entry flight and the deploying of the parachutes were textbook and the spacecraft landed in the sea just four miles from the aircraft carrier *USS Kearsarge* [CVS-33]. Within minutes a helicopter was hovering overhead and the capsule was winched out of the water. Forty minutes later the spacecraft, with Gordon Cooper still inside, was deposited on the deck of the ship. After blowing the hatch, doctors examined Gordon whilst he was still in the spacecraft to make sure that he was all right. Despite all the accolades that came his way, Gordon Cooper remained firmly with his feet on the ground, he knew that over the next few months his life was going to be one round of celebrity appearances intermingled with in-depth technical debriefings.

Among the unsung heroes, in this case heroines, at NASA, were the lady mathematicians and among them was an African-American lady by the name of Katherine Johnson. During her thirty-three year career at NACA, later NASA, she was known to be the first African-American woman to be employed as a NASA scientist. Her reputation of being able to master complex manual calculations was instrumental in the use of computers to perform the tasks. Known as the human computer, she calculated almost all the trajectories, launch windows and emergency re-entry paths for the Mercury, Gemini, Apollo and some of the Space Shuttle flights. She was also known to have worked on plans for the Mars missions. She died at the age of 101 and NASA's Computational Research Facility at Fairmont, West Virginia was named after her, such was her standing within the space community.

No more Mercury flights were planned, NASA now had sufficient information and knowledge to put into action the next phase of their space programme – Gemini.

The Saturn/Apollo tests continued with the launch on 29 January 1964 of SA-5. This was the first flight of the second stage of the rocket which went into orbit. However JPL was still carrying out attempts to photograph the surface of the Moon and on

30 January 1964, *Ranger 6* was launched. The flight was flawless and landed on the Moon as planned, unfortunately the television camera mounted on the spacecraft became damaged during the flight so no pictures were available. The next three Ranger spacecraft had their television cameras re-designed and mounted differently and were a complete success. The tests of the Saturn rockets continued with the launch on 28 May of AS-101 [SA-6] which was the first test of a boilerplate Command and Service Module.

Ranger 7 was launched on 28 July 1964 and landed on *Mare Cognitum* close to the crater *Copernicus*, sending back more than 4,300 pictures. Meanwhile the launch of AS-102 [SA-7] on the 18 September 1964 carried the first programmable-in-flight computer and was the last of the development flights. Seven months after the launch of *Ranger 7*, *Ranger 8* was launched on 17 February 1965 and deliberately crash-landed in the *Mare Tranquilitatis* area, sending back more than 7,300 pictures in the process. This was the area that was to be the site of the first manned landing on the Moon by Apollo 11 astronauts on 16 July 1969. One month later on 21 March 1965, the last of the Ranger missions was launched – *Ranger 9*. This was a complete success and crash-landed in the 75 mile [121 km] wide crater *Alphonsus*, sending back over 6,000 pictures in the process. With these encouraging results, the next stage of space exploration was about to begin.

The Apollo/Saturn tests continued with the launch on 16 February 1965 of AS-103 [SA-9], which in addition to testing a boilerplate CSM, successfully launched the first *Pegasus* micrometeorite satellite. A second satellite, *Pegasus B*, was launched on 25 May aboard AS-104 [SA-8] during another test of the boilerplate CSM. The final flight in this series of missions was carried out on 30 July on AS-105 [SA-10] with the testing of a boilerplate CSM and the launch of satellite *Pegasus C*.

One area of space travel that is rarely covered is the development of the space suits that the astronauts' lives depended on. These suits were meticulously designed and made, and were extremely expensive. The first Mercury suits, developed by the B.F. Goodrich Company and costing $7,000 each, were a derivative of the full-bodied,

The lady mathematicians of NASA. (NASA)

high-altitude pressure suits worn by USN pilots [Navy Mark. IV] after the Korean War. There were several changes incorporated into the 'Mercury' suit:

1. The open loop breathing system was replaced by a closed loop version, the oxygen entering the suit through a hose connected at the waist. This not only fed oxygen to the astronaut, but circulated it through the suit providing cooling. It exited through a hose on the side of the helmet or through the faceplate depending if it was in the open position.
2. The grey nylon outer shell of the suit was replaced by an aluminium coated nylon fabric which helped thermal control.
3. The black leather boots were replaced with aluminium coated nylon leather which helped thermal control.
4. The fitting of zips and straps to ensure a snug fit.
5. The gloves had three curved fingers for gripping the controls, whilst the middle finger was straight designed to push buttons and flip toggle switches.
6. A biometric flap was made to enable the connection of biomedical sensors to the spacecraft's telemetry systems.

Each astronaut had three tailor-made suits, one for training, one for the flight and one back-up. Of the twenty-one suits made for the Mercury astronauts, not one failed. The only suit-related incident happened at the end of Gus Grissom's flight in MR-4 when the hatch cover inadvertently blew off whilst he was in the water awaiting recovery. With water pouring in, Gus Grissom had to make an emergency exit without first securing his suit for recovery. A small inflatable life vest was fitted on all Mercury suits after this incident. There were a couple of complaints regarding poor temperature control and the difficulty in turning their heads when the suit was pressurised, but this was early days and a number of these complaints were dealt with in the design and manufacture of the later suits. These included the changes to the fitting of gloves to the suits. Alan Shepard's gloves were zippered onto the sleeves of his suit, which made rotating the wrists extremely difficult, so this was replaced by incorporating wrist bearings and ring locks on the sleeves of the spacesuit, improving mobility considerably. The black and white cloth helmet the astronauts wore, called a Snoopy Hat after the Peanuts character's aviation cap, contained all the audio equipment and were based on the type of leather helmet worn by World War Two fighter pilots. One other thing that was fitted was a UCD [Urine Collection Device]. Initially it was thought that because the flights were only to last fifteen minutes, they would not be required, but after a series of long countdown delays in which the astronauts had to relieve themselves into their thick underwear in their suits, they were fitted. There were continuous minor modifications to the suits and, on the last Mercury flight by Gordon Cooper, a mechanical seal for the faceplate was fitted which eliminated the need for the small pressure bottle and hose that had previously been used. New microphones were incorporated inside the helmet together with an oral thermometer, the latter eliminating the need for the rectal thermometer that had been used on all the previous missions.

CHAPTER FOUR

Gemini

With the Mercury and Apollo/Saturn programmes completed it was decided to get the next phase underway, the two-man Gemini programme. The name Gemini [The Twins] had been proposed by Alex P. Nagy, a NASA employee who had worked on the Vanguard project. [He received a bottle of Johnny Walker whisky for his suggestion.] The main objectives of the programme were to carry out all the trials and experiments that would become commonplace when the next phase of taking a man to the Moon, Apollo, would take place. The Gemini programme was to be instrumental in the success of all later space flights. There were several objectives in the Gemini programme: to see how long a crew could stay and carry out tasks in space, to develop a method for rendezvousing and docking with other spacecraft, and the training of astronauts in all of these procedures.

The McDonnell Aircraft Company had been selected by NASA to be the prime contractor for the building of the Mercury capsule, so in 1961 it was decided that they were best placed to build the Gemini capsule. The capsule was in essence an enlarged version of the Mercury capsule but designed to carry a two-man crew. It was 18 feet 5 inches [5.61m] long and 10 feet [3.0m] wide, with a launch weight varying from 7,100 to 8,350 pounds [3,220 to 3,790 kg]. Unlike the Mercury capsule, the retrorockets, electrical power, propulsion systems, oxygen and water, were located in a detachable Adapter Module behind the re-entry module. A major design improvement in Gemini was to locate all internal spacecraft systems in modular components, which could be independently tested and replaced when necessary, without removing or disturbing other already tested components. The instrument layout was very similar to that of the latest military fighters, something all the Gemini astronauts would be very familiar with. The astronauts usually flew with their visor on their helmets in the up position, which meant that their suits were not inflated. This made manoeuvrability relatively easy, but with the visor down and the suit, which had its own oxygen supply, inflated, the astronaut could only reach the side and bottom panels, so this is where the vital buttons, switches and handles were placed. It was also fitted with ejector seats, the standard eight ball attitude display and a radar and an inertial navigation system. The escape tower was removed, because in case of emergency in the lower atmosphere, the ejection seats would be activated. The ejector seats operated as one complete system as either astronaut could pull the ejection seat handle which was situated between their knees. After pulling the handle, the whole operation became automatic. Both hatches would be blown off and the rocket powered ejection seats activated. Just 1.1 seconds after leaving the spacecraft the two astronauts would be separated from their seats. Then 2.2 seconds later their drogue chutes would deploy, which in turn pulled the main twenty-eight foot diameter parachute out.

Access to the spacecraft was by two contoured, outwardly opening hatches with the hinges on the outboard side. The hatch was manually operated by handles both inside and outside the spacecraft. In the case of an emergency the hatch could be opened by means of an explosive device which started a three-sequence operation that unlocked the mechanical latches, opened the hatches and activated the ejection seat, all within a matter of seconds. Built into the hatches were observation windows, made up of one outer and two inner panes. The inner panes were made of toughened alumino-silicated glass, whilst the outer pane was made of a high-temperature silicate glass. The surface of the inner panes were coated to reduce reflection and glare, and to protect from ultra-violet radiation.

The heatshield that was to protect the astronauts during re-entry was of a disc-shaped design that formed the large end of the spacecraft. The ablative material that covered the heatshield consisted of two five-ply face-plates of resin-impregnated glass cloth, which were separated by a 0.65 inch [16.5 mm] thick fibreglass honeycomb core section. An additional fibreglass honeycomb structure was then bonded and filled with an ablative material called Dow-Corning DC-325 which had been developed by McDonnell. This material was poured into the honeycomb structure and immediately hardened when in contact with the atmosphere. The entire shield was then encircled with a fibreglass ring giving it greater security and covered in a material called AVCOAT and was secured in place by means of sixty-eight bolts.

Although it was described as a 'pilot's spacecraft', there were drawbacks, and one of the major ones was the cramped confines of the capsule. Gus Grissom was the smallest of all the astronauts and fitted into the Gemini spacecraft with ease, in fact it was said that he was the only one who could! All the other astronauts were literally rubbing shoulders with each other and their helmets were touching the hatches. Getting in and out of the spacecraft was difficult enough on the ground, but carrying out the same manoeuvre in space whilst carrying out an EVA [Extra Vehicular Activity] was ten times more difficult and aggravated by the fact that they were wearing an inflated pressurised spacesuit. This was a problem that was to manifest itself later and cause some worrying moments. On the extended missions of *Gemini* 5 and 7, the cramped conditions in the capsules caused painful muscle and joint problems for the astronauts and it was soon realised that the Gemini capsule was not the spacecraft to be used to take a man to the Moon. In designing the Gemini spacecraft, what is surprising is that all these problems weren't recognised and taken into consideration at the time.

Whilst there were major modifications being made to the spacecraft, there were also similar ones happening to the spacesuits. It was quickly realised that the Mercury suits were not going to be suitable for in-flight activities, which included EVAs, within the Gemini programme. The new suits were designed by NASA, based on the high-altitude suits worn by the X-15 pilots and manufactured by the David Clark

Company in Worcester, Massachusetts. Known as the G3C and G4C suits, they were to be worn by almost all the Gemini astronauts,

The G3C suit was made up of six layers of nylon and an outer layer of fire-retardant Nomex. The innermost layer was made of rubberised nylon and Nomex, giving it a bladder effect. The boots were of a combat style and made of Nomex. The full pressure helmets contained built-in headphones and microphones and the detachable gloves had improved locking rings and wrist bearings which allowed easier wrist rotation. This model was worn by the crew of *Gemini III*, Gus Grissom and John Young. All subsequent flights from *Gemini IV* to *Gemini XII* wore the G4C suit, with the exception of the crew of *Gemini VII* who, because of the length of their flight, wore an improved G4C, the G5C version. There were a number of minor modifications made to the G4C suits during this period, including replacing the Plexiglass faceplate with a high-strength polycarbonate plastic version. Another modification was the addition of layers of an insulation called Mylar that were added for aid in temperature control. On *Gemini IX-A*, Eugene Cernan's suit was modified to accommodate a test of the AMU [Astronaut Manoeuvring Unit]. Like the Mercury astronauts, the Gemini crews had three suits, Primary, Back-up and Training, all at a cost of $27,000 each.

Twelve flights were planned, ten of them to be manned. The launch vehicle was to be the Titan II rocket, built by the Martin Company. The fuel needed to launch it, although being corrosive, toxic and poisonous, did not explode in the manner of the Atlas or Saturn vehicles.

On 28 December 1961, the first ground tests of the Titan II rocket were carried out at the Vandenberg Air Force Base. It also coincided with the first launch of a Titan II ICBM [Inter-Continental Ballistic Missile] from an underground silo. The first of the Gemini spacecraft, *Gemini 1*, was unmanned, with its main objectives being to test the structural integrity of the spacecraft capsule and the modified Titan II rocket. A new tracking and communication system was installed enabling the ground support crews to train in preparation for the first manned flights. The launch of *Gemini I* had been scheduled for launch in December 1963, but like all new programmes there were teething problems and these concerned the development of both the spacecraft and the booster. Four months later, on 8 April 1964, *Gemini I* blasted off from Complex 19 at Cape Kennedy [Cape Canaveral]. The first stage was jettisoned after two and half minutes at a height of thirty-five miles and just three minutes later the spacecraft was placed into an orbit with an apogee of 170 miles [274 km].

The spacecraft stayed attached to the second stage of the rocket throughout the three orbits of the Earth, the scheduled duration of the tests. However the spacecraft completed a further sixty-one orbits until its altitude slowly decayed and it was burnt up as it entered the Earth's atmosphere. Holes had been drilled in the heatshield to ensure that it would not survive re-entry as it was not intended to be recovered.

It wasn't until the following January, that the next test flight took place, and that was a sub-orbital flight of the unmanned *Gemini II* spacecraft which was recovered. The main objective of that mission was to test the spacecraft's ablative heat protection on re-entry. There were a number of other objectives including the test results on the telemetry, fuel cells and the launch vehicle itself. The only test that malfunctioned was the one concerning the fuel cells and that went faulty just before lift-off and was deactivated by mission control. Everything else operated perfectly and the spacecraft splashed down into the Atlantic Ocean and was picked up by the aircraft carrier USS *Lake Champlain* [CV-39]. The scene was now set for the first of the manned Gemini flights, *Gemini III*, but it was a further two years before this took place.

The flight of *Gemini III* was scheduled for March 1965, but the Russians, keen to upstage the Americans, launched *Voskhod 1* in October 1964, with a three-man crew, Colonel Vladimir Komarov [Pilot], Konstantin Feoktistov [scientist and designer of the Voskhod spacecraft], Boris Yegorov [Doctor]. None of the crew wore spacesuits because there was not enough room in the capsule designed for two men to accommodate three cosmonauts wearing the bulky spacesuits. There were no ejection seats, parachutes or escape tower and the additional seating caused the crew to have to strain to read the dials, as they were still mounted in the same position as for the two-seat spacecraft. It was quite obvious that the whole mission was rushed just to upstage the Americans and to gain a political advantage for the Soviet Union. In the event that there was a major problem, it was quite obvious that there was no provision to rescue the crew: they were expendable. General Petrovich Kamanin, head of Cosmonaut training, had earlier expressed serious concerns about the mission, saying that in his opinion the spacecraft and the crew weren't ready. But such was the need to try and stay one step ahead of the Americans, that the Politburo put pressure on the programme directors to keep the missions going at any costs. Many in the Russian space programme regarded this mission as nothing more than an elaborate, politically-controlled stunt, which fortunately was completed without any major problems. A series of accidents later would highlight the mistakes made in these premature flights.

The original crew for *Gemini III* was Alan Shepard [Command Pilot] and Thomas Stafford [Pilot], but Alan Shepard developed an ear infection and was taken off the flight. They were replaced by Virgil Grissom [Command Pilot] and John Young [Pilot] whilst Tom Stafford was relegated to the back-up crew with Walter Schirra as Command Pilot. This was also the last mission to be controlled from Cape Canaveral Space Force Station, Florida, as all future space missions would be controlled from the purpose-built Manned Spacecraft Center facility at Houston, Texas.

On 28 November 1963, President Lyndon B. Johnson declared that Cape Canaveral was to be renamed Cape Kennedy after the late President John F. Kennedy, who had been assassinated on 22 November. Ten years later, in 1973, the name reverted back to Cape Canaveral, although the Kennedy Space Center retained its name.

A worldwide tracking network was established consisting of the following stations: Cape Kennedy; Grand Bahama Island; Grand Turk Island; Bermuda; Antigua; Grand Canary Island; Ascension Island; Kann, Africa; Pretoria, Africa; Tananarive, Malagasy; Canarvon, Australia; Woomera, Australia; Canton Island; Kauai, Hawaii; Point Arguello, California; Guyamas, Mexico; White Sands, New Mexico; Corpus Christi, Texas; Eglin, Florida; Wallops Island, Virginia; USNS *Coastal Sentry* [ship],

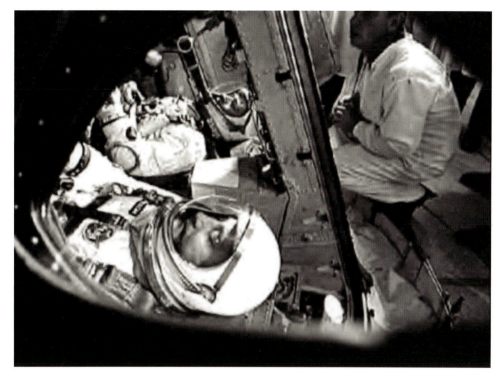

Gus Grissom and John Young in their *Gemini III* spacecraft just prior to launch. (NASA)

Quebec; USNS *Rose Knot Victor* [ship]; Goddard Flight Space Center and USNS *Range Tracker* [ship].

On 23 March 1965 *Gemini III* [*The Molly Brown*] lifted off from Cape Kennedy with astronauts Virgil Grissom and John Young aboard. John Young was the first astronaut outside of the original seven to fly in space. Their launch day had started at 5 am with the now traditional steak and eggs breakfast, followed by a pre-flight medical. They were then taken to the suiting room to be suited up in readiness for the flight. Everything ran smoothly and by 07.30 the two astronauts were inside their spacecraft with the hatches sealed. John Young found to his surprise a corned beef sandwich that Wally Schirra had smuggled on board for him. It was a friendly gesture that was to have some repercussions later.

The weather was causing some concern, as the overcast had not lifted as expected and so, thirty-five minutes before launch, *Gemini III* was put on hold. This was aggravated by the fact that the first-stage oxidiser line had sprung a leak. This was quickly solved and by this time the countdown had been resumed, and thirty-five minutes later the ground shook as the Titan II rocket lifted off the launchpad.

The crew had christened their spacecraft 'The Molly Brown', following in the tradition of the Mercury astronauts, but it was not allowed by the NASA hierarchy, who for some unknown reason thought it undignified. The second choice of name was 'Titanic', which was regarded even more unfavourably. The hierarchy relented just enough not

Gemini III launching. (NASA)

to sanction the name officially, but to accept it without further comment. Subsequent spacecraft were named by their crews, but officially they were all given Roman numerals.

Unlike the Mercury launches, the crew of *Gemini III* never felt any violent shaking during the initial stages of lift-off and relied on their instruments to keep them informed of their progress. Two and a half minutes after lift-off the first stage engine cut off and seconds later the second stage engine fired up. Three minutes later the second stage engine cut off and Grissom fired the aft thrusters to send the spacecraft into orbit. John Young, who had never been in space before, gazed in complete awe at the scene unfolding beneath him as his spacecraft started to orbit the earth. This flight was to be three low orbits of the Earth and was the first to test the manoeuvrability of the spacecraft by firing thrusters to change the shape and size of their orbit. This was to become a key factor in later flights, especially the ones destined for the Moon.

Twenty minutes into the first orbit, and just after passing over the Canary Islands tracking station, the oxygen gauge in the environmental control system showed a rapid drop in pressure. Young, whose part in the mission was to monitor all the instruments, noticed strange readings on some of the others. Assuming that there was a problem in the power supply to the instruments, John Young quickly switched the power from the primary source to the secondary source and with that all the instruments returned to normal. This decision to recognise that there was a problem, and transfer the power from the primary to the secondary source, took less than forty-five seconds. This highlighted, and endorsed, why there was such an emphasis on the intense pre-flight training that all the astronauts had to undergo.

A number of experiments were included in the mission, and one of them was for cell-growth. This unfortunately was ruined because it involved using a handle to rotate the experiment and Gus Grissom broke the handle. Whether or not the handle was too weak or Grissom used too much force is of no matter, but the experiment was a failure. The radiation experiment on the other hand was carried out successfully. It showed that the blood samples subjected to space exposure showed more damage than similar samples on the ground. The damage in fact was very small in comparison, but it did show that there was a possible danger to long exposure in space.

With the experiments completed, Grissom and Young carried out the first manoeuvre in orbit of an American spacecraft. The first Orbital Attitude Manoeuvring System (OAMS) burn lasted seventy-five seconds, which resulted in the spacecraft entering a near circular orbit of the earth and reducing its speed by 15 metres a second. Forty-five minutes later, on the second orbit, Gus Grissom fired the OAMS again and shifted the spacecraft's plane of orbit by one-fiftieth of a degree. This exercise was to test the spacecraft's transitional capability. On the third orbit, the OAMS was fired again, this time for two and half minutes, which completed the fail-safe plan, which dropped the spacecraft's perigee to 45 miles [72 km]. This manoeuvre ensured re-entry even if the retrorockets failed to ignite.

As the spacecraft approached the end of its three-orbit mission, Grissom and Young carried out their final checks and then Young fired the explosive bolts that separated the adapter from the re-entry craft. Then he fired the retrorockets that sent the re-entry module away from the adapter and the spacecraft started its descent into the atmosphere.

Everything appeared to be going well until it was realised that the planned landing point was going to be missed by over seventy kilometres. John Young knew that, theoretically, the spacecraft had a sufficient lift capability to give it some room for adjustment, but in reality there was very little. The on-board computers calculated that the landing point would be eighty-four kilometres short, and, armed with that information, John Young contacted the recovery vessels.

The parachutes opened in sequence, with the spacecraft at an attitude of 35 degrees above the horizontal but when Gus Grissom activated the landing attitude switch, the spacecraft assumed its correct attitude. The violent movement caused both astronauts to be hurled forwards in their couches, resulting in them striking their faceplates against the bulkhead. Gus Grissom's faceplate was smashed and John Young's was badly scratched.

Nestling in the water, Gus Grissom could only see water out of his observation window and realised that the parachute, which was still attached, was being pulled by the wind and was dragging them down. The memory of his ill-fated Mercury flight came flooding back to haunt him, so he quickly released the parachute, at which point the spacecraft bobbed to the surface. It was quickly ascertained that 'Molly Brown' was watertight, but the heavy swell and the increasing hot temperature in the spacecraft had its effect on Grissom. John Young, on the other hand, was a naval man and had no such problems, but Grissom refused to allow the hatch to be opened until the frogmen had arrived and attached a flotation collar around the spacecraft. After a long thirty-minute wait the welcome sound of a helicopter hovering above them heralded the arrival of the frogmen and after the flotation collar had been attached, Grissom 'cracked' the hatch and felt the cool, fresh air of the Atlantic Ocean come wafting in. Minutes later the two astronauts were winched out of the spacecraft and into the aircraft.

After medical examinations and debriefings aboard the aircraft carrier *USS Intrepid* [CV-11], the astronauts returned to Cape Kennedy. Scientists there were a bit put out regarding the experiments that had not been completed. The astronauts argued that the purpose of the mission was to carry out an engineering evaluation of the spacecraft and everything associated with it, and any time left could be used for the experiments. This time amounted to very little and the cell growth experiment was a poorly designed package [broken handle] and that they had had almost no time to work on it.

The saga of the corned beef sandwich that Wally Schirra had smuggled aboard *Gemini III* for John Young raised its head during debriefing. It was pointed out that crumbs or other parts of the sandwich could have floated behind the electronic panels and caused a problem by shorting out any of the circuits. So only food supplied by NASA would be consumed during spaceflights. The scene was now set for the next mission, *Gemini IV*.

This was to be the flight that was to carry out the first American space walk, which had been quickly added to the list of experiments and tests. The spacewalk had been brought forward because on 18 March 1965, the Russian cosmonaut Alexei Leonov had just completed Russia's first space walk from *Voskhod 2* commanded by Pavel Belyayev. He had watched as Alexei Leonov carried out his historic sixteen minute EVA, but the mission very nearly turned into a tragedy when Leonov attempted to re-enter the spacecraft. He found that his spacesuit had become rigid and he was

Gemini III splashing down in the Pacific Ocean. (NASA)

unable to move his fingers to operate the latch. This was because the difference in the air pressure between his spacesuit and the vacuum of space outside caused his suit to expand. Realising his predicament, Leonov quickly opened a valve in his suit to release some of the air, sufficient enough for him to move his fingers, open the outer hatch and clamber inside. Then followed a struggle to close and lock the hatch. This was a problem that future astronauts carrying out EVAs would become acutely aware of and know how to solve. The *Voskhod 2* mission was marred, firstly by difficulty in closing and locking the hatch after the spacewalk, and then by an onboard computer malfunction, which caused them to land 600 miles off-course in a remote area of the Ural Mountains. It was two days before rescuers could get to them. Fortunately despite being cold, tired and hungry, both cosmonauts were safe and well.

CHAPTER FIVE

The Space Race

The intense rivalry between the two countries was increasing, and missions were being planned and carried out urgently and, in some people's minds, recklessly. This was to manifest itself two years later when on 24 April 1967, the *Soyuz 1* capsule smashed into the ground after its main parachute failed to open, killing cosmonaut Colonel Vladimir Komarov. In his final moments, Komarov was heard berating the Politburo bureaucracy and the builders of the capsule for poor judgement and workmanship. There had been a number of concerns about the Soyuz spacecraft voiced by other cosmonauts prior to the mission, but had largely been ignored.

The next NASA mission '*Gemini IV*', was to be the first multi-day flight by an American spacecraft, and was intended to show that astronauts could remain and operate in a space environment for extended periods of time. With the problems suffered by Russia's *Voskhod 2* fresh in the minds of NASA, everything was being double-checked on *Gemini IV* as she was prepared for a four day, sixty-six orbit of the Earth. It was not intended to try and break the five-day record set by the Russian *Vostok 5* in June 1963. Although Russia appeared to be ahead of the Americans in the space race, they were also having serious malfunctioning problems with their spacecraft.

In America, as with previous missions, the crew of *Gemini IV*, astronauts James B. McDivitt and Edward White, were woken early in the morning, given a brief physical examination and then the traditional breakfast of steak and eggs. From there the crew were transported to the suiting area where Schmidt and his team were waiting. On the launch pad, the back-up crew of Jim Lovell and Frank Borman were testing the communication circuits and the numerous switches inside the spacecraft, so as to relieve the prime crew of this chore.

Just after 7.00 am James McDivitt and Edward White clambered into their spacecraft and prepared for lift-off. Whilst waiting in their spacecraft, Ed White's faceplate fogged up, but the suits had a built in fan system and so was quickly cleared.

The gremlins struck again just thirty-five minutes before the launch when the erector, which was lowered for the launch, stuck at an angle of twelve degrees. It was raised to its full extent and then lowered, but once again it stuck at precisely the same point. After more than an hour of searching it was discovered that a connector had been incorrectly installed in one of the junction boxes. It was re-installed and re-connected properly. At 10.16 am on 3 June 1965, television cameras recorded the launch of *Gemini IV* as it lifted off the pad and sent the

Gemini IV astronauts Jim McDivitt and Ed White in their spacecraft preparing for launch. (NASA)

images around the world. A total of 1,068 journalists from television, radio, magazines and newspapers from around the world covered the launch, all operating from a purpose built media centre.

At the point of separation from the booster rocket, Jim McDivitt turned the spacecraft around to monitor the spent vehicle. Using the thrusters on their spacecraft, McDivitt moved closer to the spent booster, but it appeared to nose down and travel on a different course to theirs. Monitoring the booster at night by means of observing the flashing lights attached to it, the crew kept in contact. As dawn approached the lights became fainter and the crew realised that the booster was some two to three kilometres away. The *Gemini IV* spacecraft was using up fuel in an attempt to catch up with the booster, so guidance was requested as to whether the planned EVA was more important than the rendezvous with the booster. The EVA was the short reply!

The problem with rendezvousing in space was not the same as it was on earth, as was later explained by engineers. Catching something on the ground relied on travelling at a speed to where the catch-up object would be at the right time. In orbit however, increasing speed also meant that the altitude was raised, which means that the spacecraft goes into a higher orbit than that of the target. This meant that the faster moving spacecraft has in fact slowed in relation to the target, because the spacecraft's orbit, which is a direct function of its distance from the centre of gravity, had increased. What should have been done, was to reduce the speed of the spacecraft, thus dropping into a lower and shorter orbit than that of the target, which would allow it to gain on the target. Then, at a pre-determined moment, when close to the target, a burst from the spacecraft's thrusters would lift it into the same orbit as the target, enabling it to close directly on to it.

The other problem that the *Gemini IV* spacecraft had, was its limited supply of fuel. By chasing the elusive booster they had used up half of their fuel supply, and could not afford to use any more in case they needed it for any fail-safe manoeuvres that may be required.

With the EVA now the priority, Ed White prepared his suit by fixing the emergency oxygen chest-pack to it and connecting the umbilical to the suit connectors. Whilst Jim McDivitt was going through the checklist with Ed White he noticed his partner was looking extremely tired. With only twenty minutes to go before they depressurised the spacecraft and cracked the hatch, McDivitt, as commander, made the decision to continue with another orbit, so as to allow Ed White to rest before going out and informed mission control of his decision.

A special 'Zip Gun' had been devised to enable the astronaut to manoeuvre in space by means of firing bursts of compressed oxygen from the gun. As they approached the end of the orbit, when Ed White was ready and had made sure his 'Zip Gun' was securely fixed to his wrist, they depressurised the cabin and started the process of opening the hatch. For a moment the hatch wouldn't open because of a broken spring, but after a bit of fiddling, the hatch swung open and Ed White released his couch harness and floated out into space.

After fixing a camera in a mount outside the spacecraft in order to capture the whole event, Ed White fired a burst from his 'gun' and gently floated above the capsule, but safely tethered by means of the umbilical cord. He then carried out a series of experiments with the gun to test the manoeuvrability it gave him. Unfortunately the compressed oxygen bottle ran out, leaving Ed White wishing that they had fitted a bigger bottle.

When he was told to return to the spacecraft, Ed White was heard to sigh and say, 'It's the saddest moment of my life.' Jim McDivitt pulled on his legs and guided them into the footwells, and after settling back into the spacecraft and storing the long umbilical, the time came to close the hatch. The difficulties they had in opening it came back – in reverse! The hatch would not lock, but after struggling for some heart-stopping minutes they managed to yank the hatch shut and seal it. This was a problem they would take up with the McDonnell engineers when they returned to Earth. The ease of opening and closing of the hatch was of paramount importance, because in the event of the crew being unable to close the hatch, the catastrophic results did not bear thinking about. The exertions required getting Ed White back aboard, and the closing of the hatch, physically exhausted the two men. They decided that this was the time to take a rest and try and get some sleep.

Jim McDivitt powered down some of the electrical systems so as to conserve power as they drifted in orbit, but because of the constant crackle of the radio as information was passed back and forth, they found it extremely difficult to sleep. This was to be their situation for the next two and half days, and the lack of sleep and restricted movement caused the astronauts to consider a series of exercises that could be carried out in the confined space of the capsule. This was an area of space travel that would have to be looked at very seriously as the need for rest and sleep was crucial.

The two astronauts carried out a series of experiments throughout the flight and also took numerous photographs with the 70 mm Hasselblad camera, as they passed over various countries. They took numerous photographs of the different weather systems that were happening below them, which gave the meteorologists a tremendous amount of information. One of the experiments they carried out was on bone demineralisation, which showed a greater mass loss in the little finger and heel than that which was experienced by long term patients in hospital beds.

A problem arose after they had completed forty-eight orbits of the earth, when they were told to turn the onboard computer off so that it could be updated. Jim McDivitt flicked the off switch only to discover that the computer would not shut down. Over the next few orbits, ground control worked frantically to try and resolve the problem, suggesting various methods to the crew. Then suddenly the computer shut itself down and quit. This meant that they would have to revert back to the type of re-entry carried out by the Mercury spacecraft. This was a rolling-type re-entry rather than the lifting bank angle that the computer would have managed for them.

During re-entry Jim McDivitt slowed the spacecraft's roll rate as they approached 89,000 feet [27,100 m] and at 40,000 feet [12,190 m] stopped it completely. Minutes later he released the drogue chute, but as it deployed the spacecraft started gyrating instead of stabilising. It wasn't until the main parachute deployed that the gyrations stopped, but it wasn't enough to stop the spacecraft slamming into the water. The landing was 50 miles [80.5 km] short of its projected landing site, but helicopters and recovery ships were soon on the scene. After 97 hours and 56 minutes in space, the crew of *Gemini IV* were home. Within minutes of the helicopters appearing, swimmers were in the water attaching a flotation collar around the spacecraft. With the spacecraft stabilised the two crew members were helped into life-rafts and minutes later were winched aboard one of the helicopters.

Once aboard the aircraft carrier *USS Wasp* [CV-7] the two astronauts were subjected to stringent medical examinations and were found to be in perfect health. The excitement of their safe return was overshadowed by the announcement that American forces were to fight alongside South Vietnamese forces in their battle against North Vietnam. This had the effect of putting the whole space programme on hold and it was two years before it was reactivated.

CHAPTER SIX

Gemini Programme Restarts

With the war in Vietnam taking up the majority of America's efforts, the Russians continued their space programme by launching the unmanned *Zond* (probe) 3 on 18 July 1965 to take photographs of the far side of the Moon in a flyby. It was also to test the transmission between the spacecraft and Earth, all of which were successful. The spacecraft was then put into a heliocentric orbit until contact was lost. *Zond 1* and *2* were launched as flybys of Venus and Mars respectively.

The American space programme got back on track and on 21 August 1965, *Gemini V*, with Pete Conrad and Gordon Cooper aboard, lifted off the launch pad. Gordon Cooper had designed a crew patch that depicted a covered wagon with a motto '*Eight Days or Bust*'. NASA's administrator James Webb, not noted for his sense of humour, refused to approve the motto because in the event of a mission failure, it would become the source of much comment. The '*Eight Days or Bust*' portion of the patch was removed.

Gemini V launching. (NASA)

The mission length was to be eight days but one of the main concerns was the supply of oxygen and hydrogen for the fuel cells. Gordon Cooper had decided to operate the cells at the lowest possible pressure in order to conserve fuel. A problem arose when Pete Conrad observed that the pressure had dropped below the operating level, and on contacting Flight Control, was told to switch on the oxygen heater in an effort to raise the pressure. It didn't work and the pressure continued to drop.

Just two and a quarter hours after lift-off, Gordon Cooper ejected the Rendezvous Pod [REP] and turned the spacecraft completely around so that radar could pick it up. The pod contained a radar transponder, flashing beacons, batteries and an antenna. The plan called for *Gemini V* to manoeuvre away from the instrument package, six miles below and fourteen miles behind, and then rendezvous with the pod. The theory was that it would drift away from the capsule and follow, but against all calculations, it went out to the side and then started to follow them.

The earlier problem with the heater was causing concern both in the spacecraft and on the ground. At the point when the spacecraft was out of communication range, Cooper decided that as he had never seen a fuel cell working at such a low pressure, he would have to power down. This of course meant that having powered down, rendezvous with the pod would now be out of the question.

Back at Mission Control a decision would have to be made whether to continue with the mission or cut it short and bring the spacecraft back. They knew there was enough power left in the fuel cells to carry out a re-entry safely, but as they debated the problem, the pressure in the fuel cell continued to drop. As the spacecraft passed on the fourth orbit, ground control were told that the fuel cells were good for at least another 13 hours. The crew of *Gemini V* also told them that the pressure had stabilised at 71 lb [32 kg] and was holding.

Cooper powered the spacecraft system's back up and to the crew's relief, the pressure continued to stabilise. Because docking with the target was no longer an option, it was decided to continue with some of the other seventeen experiments they had been tasked with. Amongst these were the tracking and photographing of celestial bodies called Celestial Radiometry, photography of the various parts of the earth as they passed over them, and a number of classified Department of Defence (DoD) experiments.

Then another problem arose when the temperature inside the spacecraft started to fall. They increased the airflow into the cabin, but this made little difference and the coolant going into their space suits seemed to be colder than usual. Unable to find the problem the crew had to put up with it as best they could.

Now that the rendezvous with the Pod was out of the question, fellow astronaut Buzz Aldrin, who had written a thesis whilst at the Massachusetts Institute of Technology [MIT], called *Guidance for Manned Orbital Rendezvous*, worked out a plan that would enable the crew of *Gemini V* to rendezvous with a 'point in space'. It worked and the first ever precision manoeuvres in space were achieved.

Throughout the flight both astronauts had been subjected to medical experiments. Some of these consisted of checking the pulse rates and blood pressures at various intervals, and calcium checks to see if there was any significant loss during the flight.

Gemini V in the Atlantic Ocean awaiting recovery. (NASA)

On the fifth day it was discovered that the OAMS (Orbital Attitude and Manoeuvring System) thrusters were malfunctioning. First one thruster quit then two more shut down, and the remaining ones became erratic. For the rest of the mission the spacecraft drifted through space, with the remaining thrusters being used only on the odd occasion to stop excessive tumbling.

On earth, Hurricane Betsy became the cause of some concern, as it was moving closer and closer to the landing area. In Flight Control, Gene Kranz, who had been keeping a wary eye on the hurricane, decided to bring the spacecraft back early and informed the recovery aircraft carrier *USS Lake Champlain* [CV-39] of the new location.

After 190 hours and 27 minutes in space, Gordon Cooper fired the retrorockets to put the spacecraft into position for re-entry. As they raced through the atmosphere, Cooper later commented, 'It was like sitting in the middle of a fire.' At 66,000 feet [20,100 m] Gordon Cooper released the drogue chute, which immediately stabilised the tumbling spacecraft. At 20,000 feet [6,100 m] the main chutes opened and the *Gemini V* gently dropped into the ocean. The spacecraft bobbed gently in the calm waters until the recovery helicopter arrived to winch them aboard and take them back to the *USS Lake Champlain* which was still 75 miles [120 km] away.

After a short break, Gordon Cooper and Pete Conrad were sent on a goodwill mission to various countries and attended the International Aeronautical Federation Congress in Athens. There they met Russian cosmonauts Aleksy Leonov and Pavel Belyayev who had flown in space aboard *Voskhod II*.

The scene was now set for the next space flights of *Gemini VI* and *Gemini VII*. This was to be the first manned rendezvous in space between two spacecraft. The idea for this had come about after the loss of the *Gemini VI* Agena Target Vehicle (GATV).

McDonnell spacecraft manager Walter Burke was overheard to say to his deputy John Yardley, 'Why couldn't we launch a Gemini spacecraft rather than an Agena?'

The idea was not a new one; thoughts about docking two spacecraft together in space had been mooted some time earlier as it would speed up the programme considerably. Among the stumbling points was the docking adapter, whether or not the tracking network could handle two manned spacecraft at the same time and persuading the powers-that-be that it was a viable proposition.

After a great deal of discussion with various members of NASA it was decided that it was something that could be considered. More meetings took place and when the Flight Control Division, who handled the tracking of the spacecraft, gave their unanimous support, the challenge was on. They decided to treat one of the Gemini spacecraft as a Mercury spacecraft and programmed their computers to read it as such. With this in place the programme was stepped up and astronauts Walter Schirra and Tom Stafford, and Frank Borman and Jim Lovell went into training.

One of the proposals put forward by the astronauts was that two of the astronauts change places during an EVA in space, but it was considered too risky at this early stage and was dismissed. Another was that a list of prescribed tasks as given to previous crews, being discontinued, and tasks and experiments could be fitted in around the flight schedule, when the crew and flight controllers thought it appropriate.

Borman and Lovell's flight was scheduled for two weeks, the longest flight to date, and storing sufficient food and gear was a problem. It was the additional storage for the expended items that caused the most problems. The space behind the two seats resolved the problem for waste matter, and the crew practised the stowing of the expendable items.

The design of a 'soft' space suit was produced with the added bonus of ease of removal. This enabled the crew to dispense with wearing the pressure suits whilst in orbit, making it considerably more comfortable for them in the confined space of the capsule.

On 4 December 1965, *Gemini VII* blasted off the launch pad and eight minutes later was in orbit 160 kilometres [99.5 miles] above the earth. Once in orbit the spacecraft was cut loose from its launch vehicle and then turned around so that the crew could monitor it. Medical experiments began within half an hour of going into a station keeping orbit, with cardiovascular cuffs being fitted to Lovell's legs.

The crew decided to catnap as the chatter between the spacecraft and the ground control became less frequent. When they awoke they had a quick breakfast and then began to run through their assignments. The astronauts felt their suits becoming warmer and decided that this was the moment for one of them to take their suit off. Despite the suits having been designed for ease of removal, because of the confined space and weightless conditions, it took almost an hour for Jim Lovell to remove his suit.

Prior to the flight there had been some discussion about the removal of suits and it was agreed that one or the other, but not both, could take off their suit once in orbit. After Lovell had removed his suit he felt very comfortable, whilst Borman continued to sweat profusely despite having his suit completely unzipped. Mission Control would not sanction the removal of Borman's suit unless Lovell put his back on. Reluctantly

Gemini VII crew in their spacecraft being readied for launch. (NASA)

Lovell agreed, and one hour later Borman was sitting in comfort while Lovell went back to sweltering in his suit. After some debate back on earth, it was decided that both astronauts could remove their suits.

Gemini VI's own trip into space was far from smooth. With their launch due on Sunday 12 December, the two astronauts, Walter Schirra and Tom Stafford, had settled into their couches aboard their spacecraft and waited for the launch. At precisely 09.54 the engines roared into life as they ignited. Just 1-2 seconds later they shut down as the malfunction detection system realised something was wrong and stopped the engines. The rocket had not even lifted off the pad; had it done so, then the shutting down of the engines would have caused it to come crashing back down. In the capsule the two astronauts, both experienced test pilots, waited, their hands hovering over the D-rings that would have activated their ejector seats. But because they had felt no movement of lift they waited, then they saw the fuel pressure lowering.

When it was realised that the rocket was not going to explode, the gantry was raised and the two astronauts quickly removed. Aerospace engineers examined the engines in great detail, bringing out all the design drawings and plans. A faulty plug was discovered and it was thought that this had become detached and was the cause, but it was subsequently found that the plug had become detached after the problem.

Gemini VI launching. (NASA)

All that day and night, teams of engineers worked without breaks and without success, but then it was discovered that a dust cover had been left over the check cover of the oxidiser inlet valve. This had come about during a routine maintenance when the check valve had been removed and a dust cover placed over the hole to prevent dirt getting in – it was never removed when the valve was put back. The mission was back on track.

On 15 December 1965, *Gemini VI-A* lifted off the pad to rendezvous with *Gemini VII*. On reaching orbit the crew prepared to catch up with *Gemini VII*, trailing the other spacecraft by 1,238 miles [1,992 km]. The *Gemini VI-A* spacecraft was in a lower orbit than that of *Gemini VII*, which meant that although its perigee remained the same, its apogee was lower than that of *Gemini VII* and it was consequently faster. After one and half hours in space, Schirra activated the thrusters to increase their speed by 13 feet [4 m] per second and increased the apogee by 169 miles [272 km].

After two and half hours he activated the thrusters once again to increase the speed by an additional 62 ft [19 m] per second to the speed of the spacecraft and then raise the perigee to 140 miles [224 km]. Thirty minutes later Schirra carried out a 90-degree turn and brought *Gemini VI-A* into the same orbit as *Gemini VII*, only 300 miles [483 km] apart and closing.

Then the *VI-A* crew got a radar lock-on *Gemini VII*. Schirra activated the thrusters once again for fifty-four seconds, increasing the spacecraft's speed by an additional 43 ft [13 m] a second. Four hours into the mission, Schirra switched control of the spacecraft into automatic rendezvous mode. As the two spacecraft closed on each other, any corrections that were going to have to be made had to be so accurate, that the crews were literally holding their breath for the most part of it.

Then suddenly the two spacecraft were less than 131 ft [40 m] apart with no visible movement between them. After receiving congratulations from both ground control and their fellow astronauts in *Gemini VII*, the two crews settled down to carry out a series of manoeuvres whilst still retaining their rendezvous positions. For the next three orbits the positions of the spacecraft varied from 1 ft [0.3 m] to 295 ft [90m] during their experiments. *Gemini VII* received instruction from ground control to cease manoeuvres as their fuel tanks had dropped to just 11 per cent. *Gemini VI-A*, on the other hand, still had a 62 per cent fuel supply.

Frank Borman and Jim Lovell watched as *Gemini VII* moved away from them and they continued on their fourteen day mission. Jim Lovell became aware of the inactivity of astronaut's legs. He commented on the fact that on this mission, their legs played no part and for the two-week period they just utilised a large space. The fact that they were in a weightless environment and there was no space to exercise

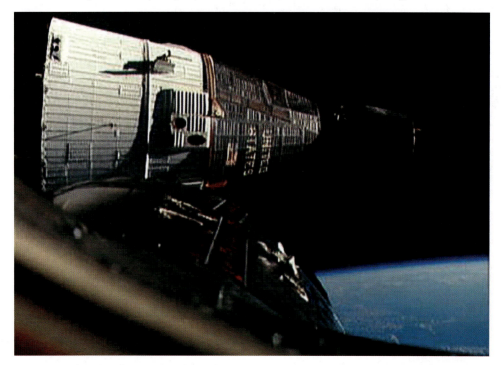

Gemini VI-A closing with Gemini VII. (NASA)

Right: Gemini
V1-A looking over
Gemini VII. (NASA)

Below: Rendezvous
in space between
Gemini V1-A and
Gemini VII. (NASA)

was a cause for a little concern. This was something that would have to be addressed when they returned to Earth.

As *Gemini VII* approached the end of its mission, concerns started to be raised regarding the fuel cells. As they approached re-entry the fuel cell warning light came on and stayed on. Lovell and Borman waited for the onboard computer to activate the retrofire switch and, to the crew's relief, they fired first time. This was followed seconds later by two more of the rockets firing. Entering the earth's atmosphere, their weightless condition suddenly left them as they became subjected to the earth's gravity, and their bodies felt as if they weighed a ton.

Then the drogue chute deployed followed a minute later by the main parachute and minutes later the spacecraft hit the water with a heavy thud. The next thing they knew was the appearance of the frogmen, who were fixing the flotation collar around the spacecraft, then came the recovery helicopters overhead to lift them out of their spacecraft and take them back to the recovery ship *USS Wasp* [CV-7].

With the success of the mission plans were immediately put into action for the next mission – *Gemini VIII*. The crew, Neil Armstrong and David Scott, had been in training for some time and they were to rendezvous with a Gemini Agena Target Vehicle [GATV] at which point David Scott would carry out a space walk to the vehicle.

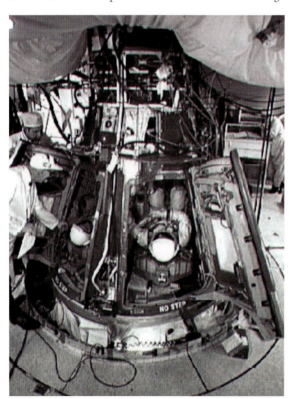

Neil Armstrong and Dave Scott in *Gemini VIII* awaiting launch. (NASA)

At 10.00 hours on 16 March 1966 the Atlas rocket roared into life and lifted the GATV off the launch pad right on time with no hitches or problems. Five minutes after launch the Agena section of the rocket took over and its engine ignited and pushed the target vehicle into its 984 mile [298 km] orbit.

Three days later Neil Armstrong and David Scott slid into their couches inside their *Gemini VIII* spacecraft and prepared for launch. At 10.40 hours the Titan II rocket burst into life and lifted the spacecraft off the launch pad. Five minutes later the spacecraft was in an elliptical orbit of 99 by 169 miles [160 by 272 km].

Making sure all the systems were up and running, the two astronauts settled down to catch the GATV. The sun set just thirty-four minutes into the flight and the spacecraft raced around the earth in the darkness. The spacecraft's thrusters cast eerie flames into the darkness, but then dawn started to break and there was some time for sightseeing. As they passed over the Pacific, Neil Armstrong commented on all the ships he could see in the Los Angeles ship basin. Then his eyes were drawn to the Rogers Dry Lake Bed at the Edwards Air Force Base where he had spent seven years as a test pilot, flying the X-15 rocket plane amongst others. Passing over Texas the two astronauts tried to spot their own homes but with no success, then it was time for a mid-course correction and other tasks.

The need to eat prompted them to each take a package of food and start to prepare themselves a meal. This was important, as on the previous flight the crews had been so busy that they had forgotten to eat and consequently found themselves tired and hungry. Weightlessness became very handy when time came for a further alignment to take place, they placed the packages above their head onto the ceiling and left them there.

The next manoeuvre came when the *Gemini VIII* spacecraft had to be pushed into the GATV's orbital path, by yawing the nose of their spacecraft 90 degrees to the south of the flight path. Then the radar locked onto the Agena at 206 miles [332 km]. One hour later Armstrong nosed the spacecraft down by 20 degrees and activated the aft thrusters, increasing the velocity to 11 miles [18 km] per second. This placed them in an orbital path just 17 miles [28 km] below the GATV.

Then 87 miles [140 km] ahead they spotted the gleaming Gemini Agena Target Vehicle and Scott switched the computer to rendezvous mode. As they approached, sunset came upon them, and as it became darker, so the gleaming Agena melted into the night. Then its acquisition lights suddenly appeared, twinkling in the darkness.

Braking the spacecraft using the thrusters, Armstrong gently eased their craft closer to the Agena until he was less than three feet away, then two feet and then they docked. One of the most difficult aspects of the space mission had been accomplished flawlessly and in the dark. With congratulation from ground control ringing in their ears, the two astronauts prepared to carry out a series of manoeuvres.

The computer aboard the Gemini Agena Target Vehicle had been designed and programmed to obey orders from both the ground and the *Gemini VIII* spacecraft. So when Scott commanded the Agena to turn them 90 degrees to the right, the command was carried out, but in less time than had been calculated. As Scott cast a test pilot's eye over the control panel, he noticed that his bank indicator showed them in a thirty degree roll. Unable to check the horizon as they were in darkness, he pointed the problem out to Armstrong. He immediately checked his instruments and confirmed that his readings were identical to Scott's.

The two locked spacecraft started to gyrate and Armstrong fired his thrusters in an effort to stop it. The movement stopped, but then four minutes later, after another command had been issued to it, the gyrations started again, only this time increasingly wildly. Despite their best efforts, the two astronauts could do nothing as the two locked spacecraft rolled violently to the left. Unable to turn the RCS off, the two astronauts became dizzy and their vision started to blur.

Ground control ordered them to disengage from Agena, but as they backed away, their spacecraft continued to whirl at a dizzying one revolution a second. As well as frantically throwing switches to try and switch off the OAMS [Orbital Attitude Manoeuvring System], Armstrong and Scott struggled with their hand-controllers. As the spacecraft continued to gyrate wildly at 17,500 miles [28,160 km] per hour, Neil Armstrong realised that the problem was a stuck thruster. The two astronauts shut off the sixteen manoeuvring system thrusters one at a time. No.8 did not respond. Then suddenly the spacecraft responded, and the spacecraft steadied. Reactivating the thrusters one by one told them that No.8 thruster had stuck open causing the spacecraft to spin violently. The spacecraft was now operating in back-up mode, but this was also the primary mode needed for re-entry.

The crew had used up 75 per cent of their re-entry manoeuvrability propellant in an effort to stop the spinning spacecraft, so ground control ordered the mission to be aborted and *Gemini VIII* to return to earth. What was concerning everybody, was whether or not the thrusters had developed leaks. If they had, the crew would have no power to get the spacecraft into the correct position for the critical retrofire that would send them into their correct re-entry path. Because the remaining part of the mission had had to be aborted, the splashdown point would have to be changed. NASA engineers worked

feverishly to ensure that all the fail-safe procedures were in place. The splashdown point was pinpointed to an area in the Pacific Ocean, 500 miles [800 km] east of Okinawa and 620 miles [1,000 km] south of Yokosuka, Japan. The destroyer *USS Leonard F. Mason* [DD-852] was dispatched at flank speed towards the area.

The *Gemini VIII* fired its retrorockets and the spacecraft passed through the atmosphere. David Scott strained to see below and then saw water just as the drogue parachute deployed, followed a minute later by the main chutes. Following the splashdown, the crew were informed that a destroyer was on its way to pick them up. In the meantime, an HC-54 from the Naha Air Base in Japan arrived on the scene and dropped three swimmers into the water close by the bobbing spacecraft.

Gemini VIII lifting off the launch pad at Cape Canaveral. (NASA)

Gemini VIII crew awaiting recovery. (NASA)

After fixing a flotation collar around the spacecraft, the two astronauts and the three swimmers waited for the destroyer to arrive. Three hours later the *USS Leonard F. Mason* pulled alongside and the two astronauts climbed a 'Jacob's ladder' up the side of the ship and onto the deck. Once they were safely aboard, their spacecraft and the swimmers were winched on to the deck.

Despite the problems to the latter part of the mission, the crew had performed the first link up in space between two spacecraft, and that was a remarkable achievement.

On 30 May 1966, the Jet Propulsion Laboratory [JPL] launched the unmanned *Surveyor 1* spacecraft to the Moon. It was to make the first soft landing on the Lunar surface in the region of the Ocean of Storms and was the first of a number of robotic spacecraft to make the journey. During its time on the surface, *Surveyor 1* transmitted 11,240 images back to Earth giving scientists the first close-up pictures of the Lunar surface.

Problems still dogged the Gemini programme, when on 17 May 1966 the unmanned Atlas launch vehicle with the Agena docking collar lifted off the pad. Ten seconds after the launch the two outboard engines were scheduled to shut down but one of them gimballed and placed the rocket in a pitch-down position and then into a nosedive position. The whole rocket was headed for the Atlantic Ocean like a torpedo and hit the water almost 125 miles [200 km] from the launch pad.

On the face of it this was disastrous, but NASA had an alternative rocket waiting and another Atlas rocket was prepared for launch on 31 May. The ATDA (Agena Target Docking Adapter) was launched on 1 June in a near perfect launch and six minutes later settled into a 185-mile [298-km] orbit. Then telemetry signals from the ATDA indicated that the shroud covering the docking port had only partially opened and had not jettisoned.

Back on the launch pad, the two *Gemini IX-A* astronauts, Tom Stafford and Eugene Cernan, were going through their countdown procedures, when the computer that was calculating their launch azimuth failed. The launch window for this mission was only 40 seconds, so with the computer shutting down, the countdown was halted and the mission scrubbed for a further 48 hours.

On 3 June at 08.40 *Gemini IX-A* lifted off the pad and six minutes later the spacecraft was inserted into orbit. On the ground, Neil Armstrong, who was the Capsule Communicator [CapCom], gave the crew the data to feed into the on-board computers for the next phase adjustments. Stafford fired the thrusters and started

Tom Stafford at the controls of *Gemini IX* during the rendezvous with the ATDA. (NASA)

after the target vehicle. The speed of the spacecraft was increased by 74.5 ft [22.7 m] per second and the perigee raised from 99 miles to 144 miles [160 km to 232 km]. As they chased after the target vehicle, the crew was busy going through the numerous checklists they had and preparing the cameras for the point of contact with the Agena.

Two and a half hours into the flight a second course correction took place: the thrusters were again fired to add another 53.15 ft [16.2 m] per second to the closing speed and the perigee raised to 171 miles [276 km]. The *Gemini IX-A* orbit was now 170 by 172 miles [274 by 276 km], which placed them 125 miles [201 km] behind and 13.7 miles [22 km] below the target vehicle and closing at a speed of 24 miles [38 km] per second.

Then the crew saw the GATV momentarily as it flickered in and out of their sight, but as they closed on it the 'running' lights on the vehicle became clearer. Stafford used the thrusters to slow their spacecraft in readiness for the rendezvous, but then as they got nearer, they realised that the ATDA (Agena Target Docking Adaptor) shroud that protected the docking adaptor had only partially released and was still attached. As the spacecraft drew alongside the two astronauts considered nudging the shroud to see if they could break it free, but on the ground the controllers said it was too risky. Stafford's description was that it looked like an angry alligator out there rotating around.

The crew rotated their spacecraft around the ATDA trying to see if there was a way of breaking the shroud free. This exercise was also one of their DoD [Department of Defence] mission objectives, finding and examining unidentified satellites. The crew determined that the explosive bolts had indeed fired, but two neatly taped lanyard wires, each of which had very high tensile strength, were holding the shroud.

Back on the ground Jim McDivitt and David Scott were sent to the Douglas plant to examine a duplicate target vehicle that was being constructed, to see if there was a way of cutting the wires holding the shroud. They determined that it could be done by one of the astronauts carrying out an EVA, but there were a number of sharp edges within the vicinity of the wires and it would be fatal if the astronaut's suit was cut. It was decided to abandon the mission in light of this, as the risk was not worth a man's life.

The cause of the shroud not releasing was later discovered to be one of neatness and not of poor workmanship. Douglas, who had built the shroud, sent it to Lockheed to be fitted to the Agena rocket's second stage. The Air Force, who were heading up the programme, then decided that the Atlas rocket did not need the second stage and that the ATDA and its fairing could be installed directly onto the rocket. This meant that the Lockheed crew were not required, and any work needed could be done by the McDonnell crew. Douglas engineers, who had been contracted by NASA to inspect and sign off on the fairing installation on the Agena rocket second stage, were also deemed to be no longer required. McDonnell engineers then installed the fairing and refused to allow any Douglas engineers to be present, despite the fact that they were completely unfamiliar with the installation. Both NASA and Douglas voiced objections, to which McDonnell said that the installation was a simple procedure and that they required no help. On the day of the launch the McDonnell crew who had installed the fairing rechecked the installation procedure laid down by Lockheed, which had been copied from a Douglas document. On the document it stated 'See Blueprint', but that document was not there. Although McDonnell had built the ATDA, it was a Douglas engineer who had supervised a practice run of the installation. The Douglas engineer who was normally involved in the process of hooking up the lanyards and knew exactly how to deal with the loose ends, was not allowed on the gantry, so the McDonnell crew fixing the shroud saw the loose ends of the lanyards and securely taped them under the small fairings that protected the explosive bolts. The lanyards in question operated the electrical disconnect to the explosive bolts and these were not connected prior to the mission for reasons of safety. It was their handiwork that had caused the problem!

With the docking no longer an option, it was decided to carry out a series of station-keeping exercises by moving into a different orbit and then re-establishing contact with the Agena some hours later using co-ordinates supplied by the ground controllers and by using their own calculations.

Twenty-one hours after launch and three rendezvous later, the two spacecraft were once again side by side. Ground control decided that this was the time for Eugene Cernan to carry out his EVA. But Stafford wanted to be well away from the ATDA before Cernan left the spacecraft, so it was decided to postpone the EVA until the end of the third day. Stafford pulled the spacecraft away from Agena and fired the thrusters, sending *Gemini VIII* into a higher orbit.

After forty-five hours into the mission, and living in the cramped confines of the *Gemini IX* spacecraft, Eugene Cernan made preparations to carry out an EVA. The process of getting into his space suit took a great deal longer than either of them had anticipated. The cramped conditions inside the capsule were the biggest problem, because the astronauts had to be careful not to hit any of the switches on the control panels. Just before sunrise on 6 June, Cernan cracked the hatch and looked out into the blackness of space. Easing himself out of his seat he floated out into the void, taking with him a litter bag which he deposited to orbit the earth and ultimately burn up.

The EVA was something he had never experienced before, not even in training. Every movement tended to send his body out of control, as for every action exacted a reaction from his body [Newton's Third Law of Motion]. Even the slightest movement would cause his whole body to move. Ed White, on his space walk, had made reference to this and stated afterwards the need for handholds and Velcro to be fitted to the outside of the spacecraft. But even these aids were not enough and Cernan found himself floating away from the spacecraft, attached only by his umbilical cord.

Making his way to the adapter he began checking the AMU [Astronaut Manoeuvring Unit] in preparation for a flight. He checked the unit's nitrogen and oxygen shut-off valves, restraint harness, sidearm controllers, and attached the AMU's tether to the spacecraft. The exertion required to carry out these relatively simple tasks was hampered by the fact he had great difficulty in controlling his body in the weightless conditions. Then his faceplate started to fog up from the exertions, causing more problems, so he decided to rest for a while to allow it to clear.

Eugene Cernan on his EVA looking back at *Gemini IX*. (NASA)

Resuming his tasks after a short break, he found his faceplate fogging up again, but he continued until he had virtually completed all his tasks and then rested again. This time he settled himself into the AMU, experiencing the most comfort he had had since leaving the spacecraft. When the decision to take the AMU away from the spacecraft and try it out came, Cernan was concerned because his faceplate fogging up was limiting his vision to a great extent. If he tried out the unit and then when the time came to take it off, he would be in free space. This meant that he would have to take it off with one hand, whilst holding on to the spacecraft with the other, and with his very limited vision, he concluded this would be dangerous. Ton Stafford agreed and contacted Mission Control who also agreed and the EVA was cancelled. Despite rubbing his nose against the fogged-up visor, which gave him a small peephole which to peer through, it quickly fogged up again. Now Eugene Cernan had to grope his way

Gemini IX crew with divers waiting to be picked up by the aircraft carrier USS Wasp in the background. (NASA)

blindly back along the side of the spacecraft to the hatch. By the time he reached the hatch, Cernan could see nothing and Stafford had to guide him into his seat. Once inside Stafford closed the hatch and pressurised the cabin.

The crew settled down to carry out the large number of experiments that had been assigned to their mission. One of these was a bioassay of body fluids, which required wastes to be collected and labelled as in a laboratory. This was an experimental requirement of all Gemini crews and one that they all disliked intensely in doing.

A second EVA was more successful and a number of DoD experiments were carried out. With almost all their experiments completed, only four remained – all scientific. Two of these were for the collection of micrometeorites and the other two were photographic, Zodiac light photography and airglow horizon photography. The micrometeorite package was attached to the ATDA and was not recovered.

The last two caused a problem, because the fogging up of the faceplate occurred again. Using the camera required being able to point it in the right direction and if you cannot see, this couldn't be done. A number of shots were taken before Cernan's faceplate fogged up and the remaining shots were taken after he had returned to the cabin.

With all the experiments completed the crew prepared for re-entry. The re-entry and splashdown were text book and 72 hours 20 minutes after launch the spacecraft floated gently in the Atlantic Ocean close by the aircraft carrier *USS Wasp* [CV-7]. The two astronauts stayed with their spacecraft whilst it was hoisted aboard.

The postflight briefing concluded that there were far too many experiments being carried out and it was a case of 'one step at a time'. The problem with the fogging

MANNED AND UNMANNED FLIGHTS TO THE MOON

up of the faceplate was amongst a number of concerns that were going to have to be addressed. The problem with overheating was finally resolved in 1989 with the development of the Apollo spacesuit together with a water-cooled undergarment which regulated the astronaut's body temperature. In a technical note by NASA the following conclusion was reached:

The ventilation system provides for two modes of operation, EV and IV. In the EV mode, all inlet gas flow is directed to the helmet for respiration and helmet defogging. The gas flow then travels over the body to the extremities, where return ducting routes the flow to the suit outlet. In the IV mode, the gas flow is split, with part of the gas flow going into a torso duct and directly over the body and the remaining gas going to the helmet. Exhaled air was returned by umbilical to the cabin environmental system, or (if on an EVA) the portable life-support system. In either case, the carbon dioxide scrubbers also removed moisture from the air. After oxygen was added, dry air was returned to the suit. Apollo spacesuits also had liquid cooling undergarments, which were very effective in preventing sweating.

The lessons learned from the *Gemini IX* flight were incorporated into the mission profile of the next mission – *Gemini X*.

The crew of *Gemini X*, Michael Collins and John Young, lifted off the pad at 15.39 on 18 July 1966 to rendezvous with their own Agena and then go into a parking orbit and rendezvous with the *Gemini 8* Agena, which still had the micrometeorite experiment attached to it. This was one part of the mission that it was hoped would be successful.

The Agena was lifted into orbit right on time and 1 hour and 20 minutes later, *Gemini X* lifted off. After settling into orbit, the crew were told that they were 1,118 miles [1,800 km] behind their Agena and their co-ordinates had been set with the computers for them to rendezvous on the fourth orbit. Using a sextant, Collins looked for the horizon but realised that he had mistaken the airglow [a band of light from the upper atmosphere] as the horizon. Unable to get a proper 'fix' they checked their figures with Mission Control and found that their figures did not match those of the computers on the ground.

After a great deal of checking it was decided that the crew should use the figures from the ground computer. Using these, John Young fired the thrusters to change their orbit to 165 by 169 miles [265 by 272 km]. This manoeuvre unknowingly turned the spacecraft slightly, causing the need for two more mid-course corrections before settling the spacecraft on the correct path. Two more course corrections later found them lined up with their Agena and they safely docked.

The course of events would have led to more practice of undocking and the re-docking, but too much fuel had been used during the lining up and course corrections. Three times more fuel had been used than in the previous mission. Despite this it was decided to go for a second rendezvous with the *Gemini VIII* Agena, so their Agena was fired to send them into a higher orbit. This placed the *Gemini X*

Gemini X on the launch pad being prepared for launch. (NASA)

spacecraft into a height never achieved before, an apogee of 474 miles [763 km] and a perigee of 183 miles [294 km]. The crew took numerous photographs from different heights as their spacecraft moved farther and farther away from earth, giving a new and interesting aspect to the view that was unfolding below them. But it wasn't all sightseeing: the crew had a large number of experiments to carry out, the large majority of them devoted to radiation and the effects of working in space. This of course was in addition to the everyday normal 'housekeeping' duties and instrument checks that had to be performed.

After nine hours of intensive work, John Young and Michael Collins settled down to catch up on some urgently needed food and rest. Nine hours later the time came to start the rendezvous sequence. Mission Control gave the crew the co-ordinates and Young made a 78-second burn to reduce the spacecraft's speed and lower the apogee. The final burn came soon afterwards and *Gemini X* lined up to rendezvous with the other Agena.

With this done the crew released their spacecraft from the Agena and Michael Collins got prepared to carry out his first experience of outer space. This was to be a stand up EVA and after cracking the hatch, Collins stood on his seat and carried out a number of experiments using a 70 mm camera, taking photographs of the southern end of the Milky Way. It was during this period that both Collins and Young began to suffer from their eyes beginning to water. Initially it was thought that it had something to do with the anti-fog compound inside their faceplates, but then it was decided that it was the result of having both suit-cooling fans on at the same time. They each switched one off and this seemed to solve the problem.

Agena rocket firing up its engine to move to a higher orbit. (NASA)

Two more mid-course corrections brought *Gemini X* within sight of *Gemini VIII*'s Agena and John Young closed their spacecraft to within 6 ft [2 m]. Young then informed Mission Control they intended to go and take a look at the micrometeorite package attached to the Agena. This was the package that Eugene Cernan had attempted to get, but because of problems with his faceplate fogging up, was unable to do so. Mission Control came back saying that Michael Collins had permission to go for an EVA, to which John Young replied, 'I'm glad about that because Mike's going out right now.' Collins picked up an experimental package and drifted out into space towards the Agena. On reaching it he discovered that there were not enough handholds and trying to hold on to smooth surfaces with the suit gloves was extremely difficult. He worked his way along to the package and by using the 'gun' manoeuvred himself to it and removed it.

Now holding two packages and the 'gun', Collins then propelled himself back to the spacecraft and then to the rear of their Agena vehicle. Only held by the umbilical cord, Collins reached the rear of the Agena adapter where he was to replace the experimental S-10 package with the one he had brought with him from the spacecraft. The problem was, that if he released the one attached and tried to replace it, he might lose one or even both of the packages, so he just removed the experiment and did not replace it.

Using the umbilical as a rope, he pulled himself back to the hatch and handed the experimental packages to John Young, before making preparations to get back inside the spacecraft. This became a more difficult problem than he had envisaged. He had become entangled in the floating umbilical and it was left to John Young to help untangle him and guide him back into his seat.

Mission Control was now becoming concerned about the amount of fuel being consumed to keep the spacecraft on station and told the crew to shut down all unnecessary equipment. With the umbilical cord snaking around inside the cabin like a giant boa constrictor the two astronauts prepared to jettison it into space. The hatch opened with ease and the cord was tossed out into space.

Re-entry co-ordinates were programmed into the computers and after ticking off the last of the experiments, the crew prepared to go home. Everything went like clockwork and the spacecraft splashed down just 3.4 miles [5.5 km] from the initial landing point. The two astronauts were lifted out by helicopter and flown back to the recovery ship, the *USS Guadalcanal* [LPH-7].

The experiments carried out by the mission with regard to rendezvousing and linking up with another, and acting as a 'space tug', were amongst the most important. This in fact was the catalyst upon which future programmes such as travelling to space stations would be based.

Whilst the Gemini missions were being carried out, another mission took place on 10 August 1966 when *Lunar Orbiter 1* was launched from Cape Canaveral. The spacecraft had been designed primarily to find and photograph smooth areas on the Lunar landscape that could be suitable landing sites for the Surveyor and Apollo missions. In addition it was to measure radiation intensity and collect micrometeoroid impact data. From 19 August until 29 August a total

John Young being helped from *Gemini X* spacecraft by a diver. (NASA)

of 42 high-resolution and 187 medium-resolution photographs were taken and transmitted back to Earth. It also took the first two photographs of the Earth as seen from the Moon. The spacecraft continued to transmit data back to Earth until it ran out of altitude control gas and impacted the Lunar surface.

The Gemini programme had gone better than hoped. There had been a number of problems, but none that weren't solvable and the information gained was invaluable. Already plans were being drawn up for the next series of manned space missions – the Apollo missions.

The next mission in the Gemini programme was *Gemini XI*, crewed by Charles 'Pete' Conrad and Richard Gordon. This was to be a controversial mission inasmuch as there were a number of projects that Pete Conrad himself wanted to do. Among these was the taking of the Gemini spacecraft to the Moon. This had been something that had been mooted some years previously when the Gemini programme had been in its infancy, but had been disregarded because plans were already in the pipeline for the Apollo missions to carry this programme out.

Another of Pete Conrad's ideas was to take the *Gemini X1* spacecraft farther away from earth than any other Gemini spacecraft had been. His reasons were that although the US Weather Bureau had satellites flying high, the photographs they returned to earth were of poor resolution. Conrad argued that the Bureau had been debating using colour photography to improve the resolution, and it would be an ideal opportunity to test his idea by photographing weather situations from even higher positions than that of the satellites.

One of the disagreements voiced against this was the fact that the Van Allen radiation belts that circle the earth might interfere with the emulsion used in the film, but a way was worked out to avoid this and so that part of the programme was given the green light.

Another suggestion put forward was that *Gemini XI* would rendezvous with its Agena on the third orbit in order to simulate an Apollo operation. But it was decided to leave the rendezvous to the first direct opportunity rather than stipulate a precise time.

One of the most important experiments was the one to try and create an artificial gravity whilst in space. One of the ways to do this was to tether the spacecraft to the Agena by means of a 118 ft [36 m] Dacron line, and to spin both vehicles at the rate of no more than ten degrees per second. Moving around in space during EVAs was the source of some concern amongst the astronauts. In order to prepare astronauts for this a programme was set up in a giant water tank. The astronaut was placed under the water until zero buoyancy was achieved, leaving him to carry out tasks that were similar to those he would have to carry out in space.

Eugene Cernan carried out a number of tests and found that the conditions were almost identical to those when working in space. But once again the most important aspect of moving around outside the spacecraft was the need for additional handholds and foot restraints. The length of the umbilical cord was also a bone of contention by both John Young and Michael Collins and the length was reduced from fifteen to nine metres.

The launch of *Gemini XI* was scheduled for 9 September 1966, but it was postponed after a pinhole leak was found in a first stage oxidiser tank. With this fixed the

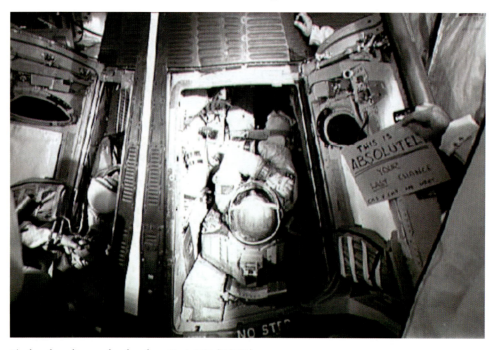

The hatches about to be closed on *Gemini XI*. (NASA)

Above left: GATV being launched prior to making a rendezvous with *Gemini XI*. (NASA)

Above right: *Gemini XI* launching from Cape Canaveral. (NASA)

launch was re-scheduled for the next day. Then faulty readings were coming from the autopilot on the GATV, so the countdown was stopped and the launch re-scheduled for 12 September.

On the morning of 12 September the Atlas powered Agena lifted off the pad and into orbit. One hour and 37 minutes later *Gemini XI* lifted off on this crucial mission. Crucial because the aim was to carry out a first orbit rendezvous, and to do this the launch window was a mere two seconds. This meant that to catch the GATV on the first orbit, the *Gemini XI* spacecraft would have to launch almost as the GATV passed overhead.

By the time the *Gemini XI* spacecraft entered orbit, the GATV was some 267 miles [430 km] away. Using calculations given to them by Mission Control, the crew performed an Insert Adjust Velocity Routine (IVAR) which not only adjusted the speed of their spacecraft, but also altered their flight path in relation to the GATV.

There was a second IVAR, but this time, because the spacecraft was outside telemetry and communication range, the calculations would have to be made by the crew and their on-board computers.

Turning on the rendezvous radar, the crew were relieved to see the lock-on signal come on immediately. As the spacecraft narrowed the gap between them, more calculations were fed into the computer together with calculations from the ground, which had come into communications range. Suddenly the blinking lights on the GATV appeared in the distance, followed almost immediately by the glare of the Sun as it appeared over the horizon in front of them.

Just eighty-five minutes after launching, *Gemini XI* rendezvoused with the GATV. The crew was then told to go for docking with the vessel, which was carried out without a problem. Because the rendezvous had been so soon after launch, the spacecraft had used very little fuel compared with previous flights. This then was the opportunity to carry out a number of docking and undocking procedures, which were all carried out by both the pilot and co-pilot in daylight and in darkness.

While this was being done a number of other experiments were being carried out. These included the study of nuclear emulsion that was in a package situated behind the pilot's hatch. With the docking procedures over, Conrad pointed the spacecraft 90 degrees away from their flight path and fired the main engine; this sent the spacecraft into a new orbit.

After resting, during which time they had some food and slept for a while, Gordon prepared for the first of his EVAs. Using information they had received from fellow astronauts about suiting up in the confinement of the capsule, they started preparing four hours before the hatch was to be opened. This, they were to discover later, was a mistake as it only took them 50 minutes to get suited up. This meant that they would be sitting in their spacecraft for the next three hours wearing the cumbersome suits.

After an hour it was decided to hook up Gordon's environmental support system and carry out tests. This unfortunately dumped excess oxygen into the cabin that had to be vented into space. This was a waste of oxygen that the crew could ill-afford. Gordon switched back to the spacecraft's system gratefully as the suit's heat exchanger had been designed to operate in space and not in the confines of a space capsule.

When the moment arrived, Gordon cracked the hatch and floated into the vacuum of space, together with a couple of garbage bags. His first project was to recover the nuclear emulsion package that he passed to Conrad. Gordon then proceeded to go forward to the docking adapter where he removed the spacecraft's docking bar and then straddled the spacecraft's nose like he was riding a horse. It took all of six minutes to fix to secure the line in preparation for the tethered flight experiment.

Conrad had been watching his colleague intensely during this time and he quickly realised that Gordon was becoming increasingly laboured in his movements. Carrying out these experiments on the ground in the simulators had been relatively simple, but here in space it was a totally different game. Conrad ordered Gordon back into the spacecraft and helped him back into his seat. With the hatch closed, Gordon took off his helmet and Conrad could see the sweat pouring down Gordon's face and into his eyes from all the exertion. Gordon was exhausted, so it was decided to cancel all further EVAs until he had rested. The result was that a number of experiments had to be aborted including the evaluation of using a power tool in space. It has to be remembered that every mission and every EVA experiment was a learning curve to the astronauts. Carrying out these experiments on Earth under controlled environmental conditions was not the same as carrying them out in space.

It was decided to abort any other EVAs with the exception of the stand-up one. It was during this one that they dumped the umbilical cord and EVA equipment into space. It was decided to go on with the next project and suited up, stowing away all unnecessary equipment.

On orbit twenty-six, Pete Conrad fired up the main engine for 26 seconds, adding 919 feet [280 m] a second to their speed and causing both astronauts to be thrust forward into their seat harnesses with the acceleration. As the spacecraft raced outwards, Gordon was constantly taking photographs and Conrad was giving Mission Control a running commentary of the view of the earth that was unfolding before him. He reported that he could see the earth, around the top for about 150 degrees and could take in Africa, India and Australia all at the same time and with no loss of colour or clarity.

The spacecraft levelled off at 853 miles [1,372 km] and whilst on station the crew took over 300 photographs. They stayed at this altitude for two further orbits and then lowered the apogee to 189 miles [304 km] where they prepared for the next EVA, the stand-up one. As they passed over Madagascar, Conrad cracked the hatch and the two astronauts, tethered by restraining straps, stood up and photographed the earth as it passed under them. The stabilization of their Agena was somewhat erratic, but not enough to cause any serious problems.

There was one major project still to be completed and that was the tethered vehicle exercise. There were two methods of completing this, the first was to dock with the GATV, point the engine section of the Agena towards the earth, then back the spacecraft out of the docking adapter until the 98-foot [30-metre] tether became taut. The two spacecraft would then drift round the earth keeping the same position and altitude. This was tried and successful to a point, but then a problem with the stabilisation of the Agena caused this method to be abandoned.

The second of the methods was to stay connected in the same manner as method one, but then by firing the spacecraft's thrusters, induce a one-degree per second rotation of both vehicles. They would then drift around the earth doing a slow cartwheel and the centrifugal force, it was hoped, would keep the tether taut and the two vehicles apart. From the start there were problems in keeping the tether taut and preventing it from snagging as the two vehicles drifted in space.

The rotation role was increased to 38 degrees per minute, which seemed to steady both vehicles. Then instructions came to increase the spin-up rate but that

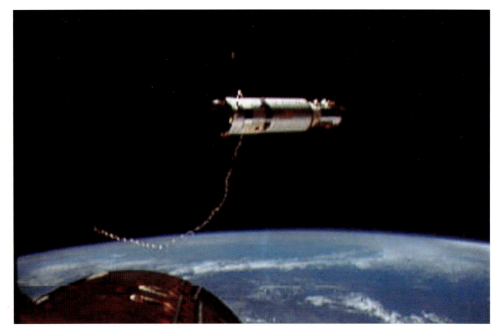

Above: *Gemini XI* carrying out tethering tests with GATV. (NASA)

Below: GATV seen from *Gemini XI*. (NASA)

MANNED AND UNMANNED FLIGHTS TO THE MOON

caused the tether to go slack and then go tight. This created what Pete Conrad called, 'the big sling shot effect' as their spacecraft was see-sawed in a pitch up to sixty degrees. Conrad, in desperation, used the spacecraft's controls in an attempt to stabilise the situation and reduce the rotation rate to fifty-five degrees per minute – it worked.

It was decided to test the micro-artificial gravity theory by placing a camera against the instrument panel and letting it go. It moved across the cockpit to the rear in a straight line and parallel to the tether. The exercise was ended when the crew released the tether and jettisoned the docking bar. With this part complete, Mission Control asked the astronauts to carry out one further test, and that was to follow the Agena by keeping to its exact orbital path. This would be a method of station-keeping albeit from a much longer distance. During the nighttime vigil, Gordon and Conrad carried out tests using night image intensification equipment by describing cities and towns in detail as they flew over them.

A second rendezvous with the GATV was carried out in just one orbit, a sign that the crew was beginning to master the technique of using the thrusters in space. Within an hour of catching up with the GATV, Conrad pulled his spacecraft alongside, before carrying out a retrograde burn to take their spacecraft away and to start preparations for return to earth. There was one more objective and that called for the crew to make an automatic re-entry.

On the forty-fourth orbit of the earth, the retro-rockets were fired. Conrad and Gordon kept all their attention focused on the on-board computer as Conrad switched from manual control to automatic. The system worked perfectly and minutes later the drogue chute was deployed followed by the main chute and their spacecraft was floating down towards the surface of the sea. On splashdown they were just over 2.5 miles [4 km]s from the recovery ship the cruiser *USS Guam* [CB-2]. The mission was a successful one and many lessons were learned.

On 20 September 1966 JPL launched the second of their proposed seven Surveyor spacecraft, *Surveyor 2*, launched from Cape Kennedy aboard an Atlas-Centaur rocket. However on its way to the Moon one of the three thrusters failed to ignite, sending the spacecraft into a spin. The out of control *Surveyor 2* crashed into the Lunar surface and was lost.

Meanwhile *Lunar Orbiter 2* was launched from Cape Kennedy on 6 November 1966. Its mission was to stereo-photograph primary and secondary landing sites, both near and far side, for Surveyor and Apollo missions. Despite a problem in the camera, 211 high-resolution images were transmitted to Earth. The spacecraft finally crashed onto the Lunar surface when the attitude control gas ran out.

Preparations were now underway for the last of the Gemini missions – *Gemini XII*. Crewed by experienced astronauts Buzz Aldrin and James Lovell, the launch took place on 11 November 1966, just five and a half hours after their GATV had been launched.

When Lovell and Aldrin walked up the ramp to the elevator that would take them to their spacecraft, they each had a card on their backs that said, 'THE END', to

Gemini XI about to splashdown in the Atlantic Ocean. (NASA)

Luna 2. This was the first spacecraft to land on the Moon. (NASA)

signify the last of the Gemini flights. Even more poignant was the fact that within hours of *Gemini XII* launching, the launch stand was being dismantled as so much scrap metal. The Apollo programme was being prepared, and five days earlier *Lunar Orbiter II* had been launched to photograph possible landing sites on the Moon.

Above: Jim Lovell and Buzz Aldrin of *Gemini XII* making their point as they make their way to the launch pad. (NASA)

Left: *Gemini XII* lifting off the launch pad at Cape Canaveral. (NASA)

Gemini XII spacecraft photographed by Buzz Aldrin during his EVA. (NASA)

Within an hour of *Gemini XII* launching, Buzz Aldrin reported back that they had a solid lock-on with their radar and they closed on the GATV. When they were within 75 miles [120 km] of the target vehicle, the reception on the radar became worthless and the on-board computers refused to accept any of the readings. This meant that the crew had to revert to the backup charts they carried. Fortunately, Buzz Aldrin had been a member of the team who had developed the chart procedures and was able to use them to great effect. Using a sextant, Buzz Aldrin measured the angle between the local horizontal of the spacecraft and the Agena. With this information he checked it against his charts and put the figures into the on-board computer.

Using the figures given back by the computer, Jim Lovell flew the spacecraft to rendezvous with the GATV and just 4 hours and 13 minutes after launching, he proudly announced that *Gemini XII* had docked. This was a tribute to the navigational skills of Buzz Aldrin and the skill of controlling the spacecraft by Jim Lovell, a perfect example of teamwork.

This was followed by a number of undocking and docking trials, but when Lovell carried out a docking manoeuvre during the night, he miscalculated the alignment and locked on just one of the latches. When he tried to undock, the latch would not release. Using the forward and aft thrusters he managed to rock the spacecraft free with no damage to either vehicle.

While this was going on, Mission Control had been monitoring the Agena's main engine and discovered that it had lost 6 per cent pressure in its main chamber. This meant that there was a corresponding loss in turbine speed and it was decided that the intended plan to fire the main engine and send the Agena into a higher orbit would be scrubbed.

With this part of the flight plan no longer an option, it was decided to replace it with one that had earlier been removed from the original flight plan, the photographing of a solar eclipse. This meant that they still had to get into a higher orbit, but they could use the power from the secondary propulsion system to do this.

After settling into the new orbit they prepared their cameras for the eclipse. Because they were taking the photographs from space as they were orbiting the earth, the eclipse for them would only last eight seconds. When the eclipse began the spacecraft was right on track and the crew managed to get photographs covering the entire eclipse.

With this part of the mission completed Buzz Aldrin prepared for the first of his EVAs. He prepared everything methodically, remembering all the advice he had been given from the astronauts who had carried out EVAs before him. When the time came to carry out his first stand-up EVA, he studied every movement of his hands, arms and body intensely. This, he knew, would stand him in good stead when the time came to move outside the spacecraft.

On 22 February 2019, Israel, together with SpaceIL, launched a Lunar Lander on a Falcon 9 B5 rocket called Beresheet, with the intention of carrying out a soft landing

Buzz Aldrin on his EVA. (NASA)

on the surface of the Moon. The lander's gyroscopes failed on 11 April 2019 causing the main engine to shut down, which resulted in the lander crashing on the Moon's surface and was lost.

After cracking the hatch, Aldrin stood on his seat and gazed out into the blackness. It was to take a further eight minutes before his eyes became accustomed to the dark void, and then the full beauty of space unfolded before him. After gazing at the stars and planets for a while, he set up an ultraviolet astronomical camera with which he photographed various star fields. In addition to this, a movie camera was set up and also a handrail to the target adapter cone. During this period the crew took advantage of the open hatch to dump their refuse bag into space. After retrieving a micrometeorite collection experiment from behind the hatch, Aldrin returned to the confines of the capsule to rest.

The following morning, Aldrin opened the hatch and drifted out into space, held only by the umbilical cord. Slowly and deliberately he moved hand over hand along the handrail to the nose of the docking adapter cone and connected the tether between the two vehicles. Returning to the open hatch he exchanged cameras with Jim Lovell and then went to the rear of the spacecraft adapter. Once there he placed his boots in the 'golden slippers' – these were overshoe-foot restraints – and carried out a series of body movements to see if they gave him more freedom to move about. They did.

With his feet in the restraints, Aldrin was able to let go of the handrail and carry out a series of mechanical experiments that included cutting metal, and undoing and doing up a series of nuts and bolts using a torque wrench. With all the experiments completed he returned to the open hatch of the cabin, wiped the window down in front of Lovell, then eased himself back in for a well-earned rest.

After resting Lovell and Aldrin backed their spacecraft away from the Agena to carry out the tethered experiments. But like the previous missions there was a problem keeping the tether line taut. When Jim Lovell started using the thrusters to maintain a distance between them and to attempt to pitch and yaw, it caused the two vehicles to roll and at one time both craft started oscillating about 300 degrees. Despite this the experiments concerning station-keeping were relatively successful.

On the 8 January 2024, a private consortium launched a Vulcan Centaur rocket from Cape Canaveral, Florida called Peregrine 1, to carry out an unmanned mission to land on the Moon. After a flawless launch and first stage separation from the Vulcan Centaur rocket, problems arose with the power supply. The solar panels, which powered the batteries, would not deploy properly, then a stuck valve caused a significant leak in the fuel tank which depleted the propellant. Without the fuel the spacecraft would be unable to carry out mid-course corrections and manoeuvre into position to land on the surface of the Moon. The spacecraft re-entered the Earth's atmosphere and was burnt up.

The next EVA took place on the fourth day of the mission and was a stand-up. This was used to carry out a number of photographic experiments and to discard any rubbish or no longer required equipment. Although this may sound as if they were using space as a dumping ground for their rubbish, anything they discarded would orbit the earth for a while before eventually re-entering the atmosphere and burning up.

On the fifty-ninth orbit of the earth the crew made preparations for re-entry and stowed away all unnecessary gear. It was to be an automatic re-entry and worked perfectly all the way into the atmosphere. There was one moment when a bag containing small pieces of equipment came away from the Velcro that attached it to the wall of the spacecraft, and dumped it in Jim Lovell's lap. Just prior to re-entry both astronauts had reactivated the ejection seats by re-fitting the D-rings that when pulled activated them. The D-ring was situated between their legs, and as the bag started to slip down after dropping into his lap, Jim Lovell resisted the urge to grab at it in fear of grabbing the D-ring by mistake. Had he done so, as commander he would have activated both seats and both he and Buzz Aldrin would have found themselves 'punched' out of the spacecraft and into the atmosphere.

The landing in the water was slightly harder than they had expected, but within minutes a helicopter arrived and dropped frogmen into the water alongside the spacecraft and had fitted the flotation collar. A second helicopter arrived and winched the two astronauts out of their life-raft and then took them to the aircraft carrier *USS Wasp* [CV-7].

The Gemini programme was over and heralded the start of the Apollo programme and the visit to another world.

Above left: *Gemini XII* splashing down in the Atlantic Ocean. (NASA)

Above right: Crew of *Gemini XII* being recovered with the *USS Wasp* in the background. (NASA)

CHAPTER SEVEN

Apollo

The concept of the Apollo programme started way back at the beginning of the Mercury programme. The Gemini programme was incorporated to fill in the gap between the two and when *Gemini XII* splashed down, the Apollo programme was already under way.

On 26 August 1965 spacecraft 012, protected by a cover with the words 'APOLLO ONE' printed on it in capital letters, arrived at the Kennedy Space Center from North American Aviation [Rockwell]. This was the Command Module [CM] that was to herald the start of the manned space programme that was to ultimately send man to the Moon. The vehicle that was to do this was the Saturn IB rocket, which

COMMAND MODULE 012 ARRIVAL KSC

The arrival of the Apollo 1 *spacecraft from the manufacturers, North American Aviation. (NASA)*

was composed of three sections, the Command Module [CM], Service Module [SM] and the Lunar Module [LM]. The Apollo spacecraft had two additional components: a Spacecraft LM Adapter [SLA] and a launch escape system [LES] to be used in case of an emergency.

The CM was the pressurised control centre of the spacecraft and contained the pressurised main crew cabin, crew couches, control and instrument panel, primary guidance navigation and control systems, communications, environmental control systems, batteries, heat shield, reaction control system that provided the attitude control, forward docking hatch, side hatch, five windows and the parachute recovery system. All this was packed into a space measuring 12ft 8ins [3.9m] in diameter with a volume of 218 cu.ft. [6.2m3]. This was the only section that would eventually return to Earth.

The SM was unpressurised and contained the main service propulsion engine together with the hypergolic propellant required to enable the CSM to enter and leave lunar orbit. The SM also contained a reaction control system that provided attitude control, radiators to filter out waste heat into space, fuel cells filled with oxygen and hydrogen reactants, and on the outside was mounted a high-gain antenna. The oxygen provided by the fuel cells not only provided oxygen for the astronauts, but produced water for drinking. The vast majority of the SM was taken up with the main rocket engine and propellants. The engine was designed to be capable of multiple re-starts and was used to place the spacecraft into and out of Lunar orbit as well as for mid-course corrections between the Earth and the Moon. At the end of the mission, the SM was jettisoned just prior to the spacecraft CM re-entering the Earth's atmosphere.

The crew of *Apollo 1*, Gus Grissom, Ed White and Roger Chaffee, had even designed their own crew patch with the approval of NASA's hierarchy. The plan was that this was to be the first manned flight followed by a second manned flight, *Apollo 2*, which would have been identical to that of *Apollo 1*, giving NASA the opportunity to iron out any of the problems discovered on the first flight. With that completed, *Apollo 3* would have the advanced Command and Service Module [CSM] which had been designed to dock with the Lunar Module [LM], creating a tunnel through which astronauts could transfer between the CSM and the LM. Well that was the theory, but it was soon realised that the spacecraft for *Apollo 2* was way behind schedule, which brought into question the need for a second launch or even a third. The idea of duplicating missions was dropped after *Mercury MR-4* when Gus Grissom had followed Alan Shepherd's mission in *MR-3*.

George Mueller, Head of the Office of Manned Spaceflight, cancelled *Apollo 2* and proposed that *Apollo 3* would be the first manned mission carrying the CSM and the LM to be launched by a Saturn V rocket. Originally the Lunar Module [LM] was called the Lunar Excursion Module [LEM] but the word 'Excursion' was dropped after it was considered to be frivolous, but all the astronauts, when referring to the Lunar Module, sounded like they were always using the word 'LEM'.

The *Apollo 1* crew: Gus Grissom, Roger Chaffee and Ed White. (NASA)

A number of other tests were carried out using unmanned rockets, but it was on 26 February 1966, with the launch of *Apollo-Saturn 201 (AS-201)*, that the first tests of the Saturn 1B and Block CSM that was to take men into space, was carried out. This was an unmanned flight designed to test the separation of the various stages that made up the Saturn 1B rocket. It was also to check the launch escape systems, the guidance, propulsion and electrical subsystems of the launch vehicle, the heatshield of the spacecraft's Command Module (CM) and recovery of the CM. The sub-orbital flight landed the CM in the Atlantic Ocean after the loss of propellant pressure caused the SM engine to shut down prematurely. All the test missions covered the testing of the ablative heatshield that protected the spacecraft on re-entry. The heatshield consisted of a stainless steel shell with a fibreglass honeycomb structure that had been molded to the curvature of the shield, which created a base for anchoring the ablative material to the contoured shape. This was impregnated with a phenolic resin which was bonded to the shell with an epoxy-based adhesive. All the edges of the compartments and hatches were protected by epoxy-fibreglass to prevent erosion by the shear forces the spacecraft would be subjected to during re-entry. One method of creating a heatshield for the spacecraft that was proposed, was making tiles and cementing them to the spacecraft. It was dismissed as not being practical, but interestingly enough it was this method that was used some years later on the Space Shuttle's Orbiter spacecraft.

On 5 July the second of the unmanned Apollo rockets, *Apollo-Saturn 203 (AS-203)*, was launched. This was to check out:

    a. Venting and chill-down systems,
    b. Fluid dynamics and heat transfer to propellant tanks,
    c. Attitude and thermal control systems,
    d. Launch vehicle guidance,
    e. Checkout in orbit.

The following month the *Apollo-Saturn 202 [AS-202]* was launched, to examine and evaluate the following:

    a. Command Module (CM) heatshield at a high heating load,
    b. Structural integrity and compatibility of the launch vehicle,
    c. and the spacecraft,
    d. The flight loads,
    e. Stage separation,
    f. Subsystems operations,
    g. Emergency detection system operations.

With unmanned tests concluded, preparations got underway for the first of the Apollo missions. At the beginning of the Apollo programme, the only suits available were modified Gemini ones. In 1962 NASA held a competition for the contract to design and make the Apollo Space Suit Assembly [SSA]. There were a number of proposals which were finally whittled down to just two: Hamilton Standard Division of United Aircraft Corporation making the Portable Life Support System [PLSS], with the David Clark Company providing the Pressure Garment Assembly [PGA], and the International Latex Dover Corporation [ILC] as the SSA programme manager, together with Republic Aviation and Westinghouse providing the PLSS. After much evaluation NASA chose the Hamilton Standard proposal for the PLSS and the ILC Dover proposal for the PGA, but by March 1964 the standard required had not been met. It was decided to divide the spacesuit programme into three sections, Block I, Block II and Block III. Block I would be the David Clark Company, who had made the Gemini suits, and would provide the suits for the early EVA missions. Block II, the Hamilton Standard/ILC Dover companies, would support the early EVA missions for the time being. Block III would cover the later and more advanced pressure suits and backpacks [PLSS] that would be selected from a future competition.

Problems arose when it became obvious that Hamilton Standard and ILC were unable to form an effective partnership; this then forced NASA to take over the Block II management. Professor Matthew Radnofsky, Head of Crew Equipment Branch at Houston, monitored the development of the suit. In July 1965, ILC were awarded the contract for the Block II suit with its AFL design, which included the inner garment, and together with the Hamilton Standard backpack, completed the selection of the spacesuit and backpack design. There followed a number of modifications to the suits as missions evolved culminating in the A7L. The first of these was worn by the *Apollo 7* and *Apollo 8* crews and were used as launch and re-entry suits.

With the launch of *Apollo 9* on 3 March 1969, came the new Apollo EMU suit and tests were carried out during the transition of the LM from the CSM. The tests were successful with the result that Block III was scrapped, as the modifications to the suits had reached the standards required.

The make-up of the suits was extremely complicated and were made by ILC Dover. The workforce, mainly women, usually made ladies' underwear [Playtex] and were highly skilled seamstresses, but the manufacture of the spacesuits took their skills to a new higher level as there was no room for error. They were to produce a suit that was to provide a life sustaining environment, one that would protect the astronauts during EVAs and unpressurised operations.

The astronaut's inner suit was a one piece underwear similar to a 'Long John' made of Spandex, with 300ft [91.5m] of narrow rubber tubing circulating throughout and close to the skin, through which water was pumped to keep the astronaut cool. Sweat was released through vents in the suit and was recycled into the water-cooling system. Oxygen was also pumped in at the wrists and ankles to aid the circulation in the suit.

Each suit consisted of twenty-one layers of fabric, Beta cloth, Nylon Tricot, Spandex, Neoprene, Dacron, Aluminium-coated Mylar, Kevlar and Nomex, with an outer layer of Teflon. An astronaut got into the suit through a rear pressure-sealing zipper that ran from the upper back through the crotch. The suits were designed and manufactured to withstand temperatures ranging from -25°F to +230°F. The astronauts wore gloves and helmets, the gloves were securely connected to the suits just above the wrists and the helmet at the neck, both on red metal bearing rings. All the suits were X-rayed twice to ensure that no pins were left behind by the seamstresses [none were ever found], once at ILC Dover and again at NASA. The overshoes, with their distinctive ribbed soles, were only worn on the Moon and were to be left behind on the surface along with the PLSS so as to reduce the launch weight. It is interesting to note that the A7L Apollo spacesuit with the PLSS attached weighed 81 pounds [37 kg] on Earth, but only 30 pounds [13.6 kg] on the Moon. The PLSS was 26 inches (66 cm) high, 18 inches (46 cm) wide, and 10 inches (25 cm) deep and was the major life support system for all the astronauts that walked on the Moon. Once outside the spacecraft the astronaut could walk and work in his own environment and as he breathed the oxygen from the tanks inside the backpack. Lithium hydroxide removed the carbon dioxide from the expelled air and circulated water through an open loop in the tubing in the inner garment. Also contained in the PLSS was a radio receiver and transmitter for communication, which was relayed through the spacecraft's communication back to Houston. The controls for the PLSS were mounted in the Remote Control Unit [RCU] on the chest of the astronauts. In order to see the controls, a mirror was strapped to the wrist, enabling the astronaut to see and operate the RCU. The first four Lunar surface missions [*Apollo 11* to *14*] were limited to four hours, with oxygen stored at 1,020 pounds per square inch [7.0 MPa], 3.0 pounds [1.4 kg] of lithium hydroxide, 8.5 pounds [3.9 litres] of cooling water, and a 279 watt-hour battery. For the extended missions of *Apollo 15* through *17*, the EVA stay time was doubled to eight hours by increasing oxygen to 1,430 pounds per square inch [9.9 MPa], lithium hydroxide to 3.12 pounds [1.42 kg], cooling water to 11.5 pounds [5.2 litres], and battery capacity to 390 watt-hours. In the case of an emergency, a separate unit, the Oxygen Purging System [OPS], was mounted on top of the PLSS directly behind the astronaut's helmet. The OPS maintained suit pressure and removed carbon dioxide, heat and water vapour through a continuous, one-way air flow vented to space. When activated, the OPS provided oxygen to a separate inlet on the pressure suit, once a vent valve on a separate suit outlet was manually opened. The OPS provided a maximum of about 30 minutes of emergency oxygen for breathing and cooling. This could be extended to 75 to 90 minutes with a 'buddy system' hose that used the other astronaut's functional PLSS for cooling [only]. This allowed the vent valve to be partly closed to decrease the oxygen flow rate.

All the Apollo astronauts carried a maintenance kit to enable them to carry out minor repairs to their spacesuits. These contained exterior patches, cloth tape, sealant for bladder repair, optical surface cleaning for their helmet faceplates and defogging pads. One of the minor problems that affected some of the astronauts was when wearing their full suit complete with helmet, they had the urge to scratch their nose or their ear. Now as innocuous as this may sound, it can become the source of extreme irritation and detract from the task ahead. To try and resolve this a small piece of Velcro was attached to the inside of the helmet, which enabled the astronaut to turn his head and rub his nose or ear against it – and it worked. It had taken over three years to produce the Apollo Extravehicular Mobility Unit [EMU], and the cost of the Apollo spacesuit, bearing in mind that every astronaut who flew on a mission had three (primary, back-up and training), was $250,000 each. The PLSS were priced at $250,000 each. Later, when the Space Shuttle became operational, the cost of the EVA suits, including self-contained life support systems, reached a staggering $450,000 each.

Then on 27 January 1967, tragedy struck. The three astronauts, Lieutenant Colonel Virgil Ivan 'Gus' Grissom, USAF, Lieutenant Colonel Edward White, USAF, and Lieutenant Commander Roger Chaffee, USN, were in the Command and Service Module [CSM] aboard *Apollo 1*, which was to be the first manned spacecraft of the Apollo programme. The CSM was on top of a Saturn IB (AS-204) launch rocket, which was on launch complex 37B at the Kennedy Space Center, and were going through a simulated countdown under launch conditions. Suddenly over the intercom came Roger Chaffee's voice:

'Fire!'

'We've got a fire in the cockpit!'

'We've got a bad fire... Let's get out! We're burning up...'

Seconds later this was followed by a piercing scream. It was determined later that a spark must have ignited the pure oxygen atmosphere in the capsule and within a matter of seconds the three astronauts were dead. The inside of the capsule had been turned into a raging inferno, estimated to be in excess of 1,000 degrees centigrade in less than ten seconds. The fire had been so rapid and so intense that it was some time before the rescue crews could get inside the capsule and remove the bodies of the three astronauts. Their space suits were burnt and charred and it was discovered during the autopsy that Gus Grissom's body had suffered some 60 per cent burns, whilst Ed White's had suffered 40 per cent and Roger Chaffee's 30 per cent. The cause of death given for all three astronauts was said to be due to: 'inhalation of toxic gases caused by the fire'.

The whole of America was stunned, none more so than their fellow astronauts. Immediately the question on everybody's lips was how could this have happened? Safety was paramount at all NASA complexes and facilities, and although it was accepted that accidents were bound to happen, this went beyond what was deemed to be just an accident. Astronauts Walter Cunningham, Don Eisele and Frank Borman listened to the tapes that had recorded the last minutes of the crew over and over again, but could find nothing in the tapes that would give them a clue as to what had happened. Investigations followed and lessons were quickly learned, but it was never fully established beyond doubt what actually caused the fire. Any evidence that could have determined the cause was destroyed by the ferocity of the fire. But by a process of elimination, one of the main theories put forward was that a wiring fault beneath the left-hand couch [Gus Grissom] ignited the pure oxygen atmosphere in the CM.

What exactly caused the fire may never be known, but it is generally accepted that it was a spark from faulty wiring that was to be the most likely cause. It was determined that all three astronauts were dead before the fire itself engulfed them.

There had been a great deal of criticism about the Apollo spacecraft prior to the fire and the incident did nothing but add fuel to these concerns. In earlier company tests of the main engine, the exhaust nozzle shattered when the engine was fired. In another test when the capsule was dropped during a water-landing test, the heat-shield cracked wide open. It is said that Gus Grissom, after witnessing some of the tests, was so disgusted with the results that he placed a lemon on the top of the spacecraft. During one of the transmission tests from the capsule there was a great deal of static, which resulted in a great deal of missed conversation. This caused Gus Grissom to comment that if they could not communicate with Mission Control from the top of the launch platform, what chance was there from the Moon?!

The results from the accident investigation had widespread effects, not only for NASA but for North American Aviation [Rockwell], the manufacturers of the Service Module [SM] and Command Module [CM], after discrepancies were revealed in the overall design and shoddy workmanship practices.

The main concern now was to ensure that such an accident did not happen again, and one of the major problem areas that was identified, was the fact that the ingress/egress hatch opened inwards. The hatch system consisted of three hatches. The inside hatch had a rack-drive bar that operated six latches that when activated, clamped the hatch to the walls inside the module. The next outer hatch cover consisted of one of the most complicated locking systems imaginable, and consisted of twenty-two latches that were activated by a centre lock, which in turn was operated by push-pull

The interior of *Apollo 1* after the fire that killed all three astronauts.

rods, rollers and bell cranks. On top of this was an outer cover that was released after the spacecraft was in orbit.

There were a number of changes to the Apollo spacecraft that had come from the results of the inquiry into the fire aboard *Apollo 1*. The astronauts who had been among the members of the different inquiry teams that were looking into the tragedy, demanded changes to be made. In an emergency and under perfect conditions, to open the existing hatch, it was estimated that it would take a highly trained team of astronauts, aided by a similarly trained team of technicians on the outside, more than 90 seconds to open. This was deemed to be unacceptable, and the whole thing was redesigned to become a gas operated, single hatch that opened outwards. It was also estimated that it would only take between four and seven seconds to exit the spacecraft in an emergency with the new hatch in place. A number of other alterations were also made to the design and construction of the command module before the next crews were destined to step inside it. All the wiring inside the spacecraft had to be of a determined standard and fireproof. The atmosphere inside the spacecraft whilst on the pad was to be 60-40 nitrogen/oxygen mix and all fabric surfaces, including space suits, had to be made of non-inflammable Beta cloth. With these new changes agreed and implemented, the time was felt to be right for the first manned flight to take place.

MANNED AND UNMANNED FLIGHTS TO THE MOON

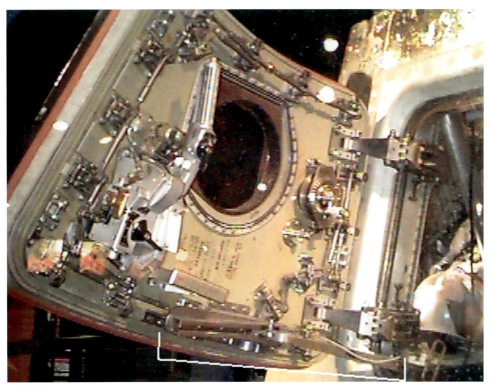

The newly designed hatch on *Apollo 7*. (NASA)

The next designated crew to fly was Walter Schirra, Don Eisele and Walter Cunningham. Schirra had previously flown on both Mercury and Gemini flights and consequently was a very experienced astronaut. To ensure that all the new safety features were being incorporated into the new command module, the three astronauts virtually haunted the facility where the module was being built at North American Rockwell, and checked the workmanship constantly, much to the annoyance of some of the construction crews. Their only answer to the critics was, 'You only have to build it, but we have to live in it, and our lives depend on it.' Walt Cunningham was the third civilian astronaut, the first two were Neil Armstrong and Elliot See. NASA also announced that the atmosphere inside the spacecraft would no longer be 100 per cent oxygen whilst on the launch pad, but would be the recommended 60 per cent oxygen and 40 per cent nitrogen. The 100 per cent oxygen atmosphere would only be retained whilst in space.

The programme continued with *AS-201*, *AS-202* and *AS-203*. These had been destined to be called *Apollo 1*, *2* and *3*, but with the accident it was decided to continue with unmanned testing flights, with *Apollo 4* being the next full Apollo flight. The launch of the unmanned *AS-201* on 26 February 1966 aboard a Saturn 1b, an uprated Saturn 1, had been the first flight of this rocket and carried a Block 1 CSM, which meant that it was unable to dock with the LM. This was the first of the many 'boilerplate' models

Above left: Unmanned *Apollo 2* launching from Cape Canaveral. (NASA)

Above right: Unmanned *Apollo 3* launching from Cape Canaveral. (NASA)

that were to be used for testing on the various test flights. The *AS-201* had also carried, for the first time, the Spacecraft Lunar Module Adaptor [SLA] and a Block 1 Launch Escape System [LES]. The 'Block 1' refers to the first design and, as information came back from the various test flights, so it was refined. Also tested and evaluated was the ablative heatshield on the CM. The *AS-203* was launched on 5 July and tested the venting and chill-down systems, the heat treatment to the propellant tanks, attitude and thermal control system, launch vehicle guidance system and the Command Module [CM] heatshield at high temperature.

The final Orbiter mission, *Lunar Orbiter 3*, took place on 5 February 1967. The mission was similar to the two previous missions and that was to stereo-photograph possible Surveyor and Apollo landing sites.

The last of the unmanned flights was *AS-202*, whose primary object was to further evaluate the CM heatshield, test the structural integrity of the rocket, the flight loads, stage separation subsystem, and the emergency detection system. With all these objectives met, the date was set for the first manned Apollo mission. Just less than a year later on 9 November 1967, the first of the unmanned Apollo flights that was given the Apollo classification was *Apollo 4 [AS-501]*, and was launched from Launch

Complex 39. This was the first flight of the Saturn V rocket and had a spacecraft attached, consisting of a Lunar Module Test Article [LMTA], built by Grumman, and a Command and Service Module [CSM] built by North American Aviation.

To launch this massive Saturn V rocket required a unique piece of equipment known as the Missile Transporter Facilities or simply the Crawler-Transporter. Two Crawler-Transporters were ordered to be built by the Marion Power Shovel Company, based upon the Bucyrus-Erie 2,700 metric ton crawler shovel, using components designed and built by Rockwell International at a cost of $14million each [$128.5million in 2022]. They were the largest self-powered land vehicles in the world and weighed a staggering 2,721 tonnes. The Crawlers, named *Hans* and *Franz*, measured 131 ft by 114 ft [40 m by 35 m] with an adjustable height from 20 ft to 26 ft [6.0 m to 7.9 m] with each side being raised or lowered independently of the other. It was moved on eight tracks, two on each corner, with each of the tracks having 57 shoes, each shoe measuring 7.5 feet long by 1.5 feet [2.3 m by 0.46 m] wide and weighing in at one ton, making a total of 456 shoes per Crawler. A team of around thirty engineers, technicians and drivers were required to operated the Crawler, which was driven from two control cabs situated at either end of the platform.

The Crawler was powered by sixteen traction motors, each powered by four 1,000 kW [1,341 hp] generators, in turn driven by two 2,050 kW [2,750 hp] V16 ALCO 2510 diesel engines. Two 750 kW [1,006 hp] generators, driven by two 794 kW [1,065 hp] engines, were used for jacking, steering, lighting, and ventilating. Two 150 kW [201 hp] generators were also available to power the Mobile Launcher Platform. The Crawler's tanks held 19,000 litres [5,000 US gal] of diesel fuel, and it

The giant 'Crawler' mobile launch pad. (NASA)

The Crawler with engineers walking alongside showing the immense size of the mobile launch pad. (NASA)

burned 296 litres per kilometre [125.7 US gal/mi]. The Crawlers travelled along the 5.5 and 6.8 km [3.4 and 4.2 miles] Crawlerways, from the Vehicle Assembly Building [VAB] to Launch Complex-39A and Launch Complex-39B, respectively, at a maximum speed of 1.6 kilometres per hour [1 mph] loaded, or 2 mph [3.2 km/h] unloaded. The average trip time from the VAB along the Crawlerway to Launch Complex 39 was about five hours. Each Crawlerway is 7 ft [2 m] deep and covered with Alabama and Tennessee river rock for its low friction properties to reduce the possibility of sparks.

JPL launched the third of their Surveyor spacecraft, *Surveyor 3*, on 17 April 1967 from Cape Kennedy. This was the first spacecraft to be fitted with a surface-soil-sampling-scoop, which dug four trenches 7 inches [18 cm] deep. The Lunar soil was then placed in front of the television camera and the images transmitted back to Earth. On the first night, the solar panels on the spacecraft were shut down as they were not producing any electrical power. The following morning the spacecraft could not be activated because the extremely cold night-time temperature had damaged the solar panels. When *Apollo 12* landed on the Moon within walking distance, Pete Conrad visited the *Surveyor 3* spacecraft and took several photographs of it as well as removing the scoop.

The Orbiter programme continued with the launch, on 4 May 1967, of *Lunar Orbiter 4*. The three previous Orbiters had been tasked with finding landing sites; this time *Orbiter 4* was to orbit the Moon and carry out a broad photographic survey of the features of the Lunar surface in order to find sites thought to be suitable for scientific exploration. Despite problems with the thermal camera door, the Orbiter transmitted 419 high-resolution and 127 medium-resolution images back to Earth and succeeded in covering 99 per cent of the near side of the Lunar surface. It crashed into the surface of the Moon six months later after going into a decaying orbit.

The fourth Surveyor spacecraft, *Surveyor 4*, blasted off from Cape Kennedy on another surveying mission of the Moon. After a flawless flight and just two and half minutes before touching down for a soft landing, all communication was lost. Despite repeated efforts to re-establish contact there was no response. It is thought that the solid propellant in the retro-rocket may have exploded, destroying the spacecraft.

The last of the Lunar Orbiter missions took place when on 1 August 1967, *Lunar Orbiter 5* was launched from Cape Kennedy. This final mission was in two parts: the primary mission was to take photographs of prime landing sites for the Apollo missions; whilst the secondary mission was to acquire the precise trajectory for improving the lunar gravitational field. In addition it was to measure the radiation and the micrometeoroid flux of the lunar environment, as well as continuing to map the Lunar surface. During its time in orbit, the spacecraft photographed thirty-six different areas of the near side of the Moon and mapped almost all of the far side, which included twenty-three previously unphotographed areas. *Lunar Orbiter 5* was crashed into the Lunar surface on 31 January 1968.

Despite the setback and the loss of *Surveyor 4*, *Surveyor 5* was launched from Cape Kennedy on 8 September 1967. After a flawless flight, and just prior to touchdown, a leak was discovered in the spacecraft's thruster system, but engineers at JPL managed to correct it and carry out a soft landing on the Lunar surface in the south-eastern region of Mare Tranquilitus. *Surveyor 5* took more than 20,000 photographs and transmitted them back to Earth in addition to carrying out a number of scientific experiments. It also carried out the first lunar soil analysis.

Surveyor 6 was launched on 7 November from Cape Kennedy, and after an uneventful flight soft landed on to the surface of the Moon. In the next two weeks it sent 29,952 images back to Earth before being put into hibernation mode fot the Lunar night. Three weeks later the spacecraft was woken up and a command sent for it to fire its three thruster rockets. The spacecraft rose about 10 ft [3 m] and moved sideways about 8 ft [2.5 m] before settling down again. This gave JPL scientists the opportunity to examine the original landing footprint to determine the soil's properties. It was then shut down for the duration.

The launch of an unmanned *Apollo 4* on 9 November 1967 was to carry out a full evaluation before the first manned flight took place. Not only was the launch vehicle to be fully evaluated, so was the spacecraft itself. This was to be one of the most intensive evaluations of the programme.

After launch the first stage cut-off occurred two minutes and thirty seconds into the flight at a height of 39 miles [63 km]. The second stage ignition occurred two seconds later and lasted six minutes and seven seconds, followed by the ignition of the S-IVB stage, giving a burn of two minutes and twenty-five seconds. This placed the S-IVB and the spacecraft into an Earth parking orbit with a perigee of 113 miles [182 km] and an apogee of 116 miles [187 km]. Three hours into the mission and after two orbits of the Earth the S-IVB stage was fired to place the spacecraft into a simulated lunar trajectory. The burn lasted five minutes and when completed the spacecraft and the S-IVB were separated.

The firing of the service module propulsion system, which placed the Command and Service Module [CSM] into an apogee of 11,344 miles [18,256 km], with the heatshield of the Command Module [CM] away from the Sun, followed this. The spacecraft was left in this position for four and half hours to test the thermal integrity of the spacecraft. After eight hours into the mission the CSM propulsion system was ignited once again and the spacecraft placed into a re-entry trajectory. The burn, which lasted four and a half minutes, caused the spacecraft to enter the Earth's atmosphere at a velocity of 6.65 miles [10.7 km] a second. All went well and the spacecraft was recovered from the Pacific Ocean off Hawaii, by the aircraft carrier *USS Bennington* [CV-20]. The mission was a complete success.

As an aside, it is interesting to note that the tanks that held the cryogenic (ultra-cold) liquid oxygen and liquid hydrogen on the Apollo spacecraft came close to being the only leak-free vessels ever built. It was estimated that if a car tire leaked at the same rate as the tanks, it would take the tire over 32,400,000 years to go flat. The fuel rate of a launch of a Saturn 5 rocket was estimated at 5 inches [12.7 cm] per gallon.

The last of the Surveyor programme, *Surveyor 7*, was launched on 7 January 1968 and was the fifth to achieve a soft landing on the lunar surface. The landing site had been chosen so as to be well away from the other Surveyor spacecraft, enabling it to look at a different terrain. *Surveyor 7* carried more scientific equipment than any of the other previous Surveyors and included a television camera with a polarising filter, a surface sampler, a surface scoop fitted with two magnets to detect any metals, and eleven auxiliary mirrors. Seven of the mirrors were positioned to show the lunar samples as they were collected by the scoop.

On 22 January 1968, the second of the unmanned Saturn IVB rockets carrying the Lunar Module [LM], *Apollo 5 [AS-204]*, blasted off from Launch Complex 37B at the Kennedy Space Center. *Apollo 5*'s mission was to verify the operation of the Lunar Module's structure itself and its two primary propulsion systems. It was also to evaluate the Lunar Module staging and the orbital performance of the S-IVB stage and instrument unit. Soon after achieving orbit, the nose cone, which replaced the CSM, was jettisoned and the LM separated. The first firing of the descent engine went virtually as planned, except that the Lunar Module's guidance system shut down after only four seconds of operation. This was because the engine's velocity did not build up

Unmanned *Apollo 5* launching from Cape Canaveral. (NASA)

as quickly as predicted. Houston pinpointed the problem in the guidance system itself and not in the hardware design. This enabled the engineers and scientists at Mission Control to pursue an alternative mission that achieved the same objective. After the mission had been completed, the following day the Lunar Module stages were left in decaying orbit to burn up on re-entry at a later date. A third unmanned flight to test the Lunar Module [LM] had been scheduled, but due to the success of the first, LM II was canceled. The LM II now stands in the National Air and Space Museum (NASM) in Washington DC.

The last of the unmanned Apollo flights, *Apollo 6 [AS-502]*, was launched from Complex 39A at the Kennedy Space Center on 4 April 1968. Just two minutes and

Unmanned *Apollo 6* launching from Cape Canaveral. (NASA)

thirteen seconds after launch, problems were observed in the second stage during the boost phase. Four minutes later two of the J-2 engines shut down early, so the firing of the remaining three engines was extended for one minute to compensate. This sent the third stage into a higher orbit than was planned, 110 by 226 miles [177 by 363 km] rather than 100 miles [161 km] near circular orbit. The amount of fuel used curtailed the mission to a large degree, and the spacecraft was returned to Earth after nine hours and fifty minutes after launch. The amphibious assault ship *USS Okinawa* [LPH-3] recovered the spacecraft.

In the USSR plans were being made for a manned flight to the Moon and the launch on 21 September 1968 of the unmanned *Zond 5* from Baikonur heralded their intent. The spacecraft was sent on an orbit around the Moon, photographing possible landing sites and then returned safely to Earth. This was the first spacecraft to orbit the Moon and return. On board were two live tortoises, who survived the mission, and a number of insects and plants. It is said that in the 'pilot's seat' was a maquette made to the size and weight of a man. Preparations then got underway for the launch of *Zond 6* with the expectation, assuming there were no technical mishaps, to plan for a manned flight.

In America, the time was now deemed right for the first manned Apollo spaceflight, *Apollo 7*. The crew, Walter Schirra [Commander], Don Eisele and Walter Cunningham, had been training continuously since they had been told that they were to be the next prime crew. So too had their back-up crew of Frank Borman, Thomas Stafford and Michael Collins.

Recovering the *Apollo 6* capsule after the last unmanned launch. (NASA)

With the thoughts of landing a man on the Moon firmly fixed in their minds, the Manned Spacecraft Center (MSC) directors had to look seriously at the spacecraft that would actually land on the Moon's surface. At Ellington Air Force Base, Texas, trials were being carried out on a Lunar Landing Research Vehicle [LLRV], but these received a setback on 6 May 1968 when NASA test pilot Neil Armstrong [*Apollo 11*] had to eject from the research vehicle after losing control. The LLRV was completely destroyed at a cost of $1.5 million.

Left: The Lunar Landing Research Vehicle [LLRV] being test flown by NASA test pilot Neil Armstrong. (NASA)

Below: Neil Armstrong parachuting to safety after losing control of the LLRV. (NASA)

CHAPTER EIGHT

Destination Moon

On the 11 October 1968, the first manned Apollo spacecraft, *Apollo 7* [AS-205], lifted off from Cape Kennedy. Astronauts Walter Schirra [Commander], Don Eisele and Walter Cunningham had climbed into *Apollo 7*'s Command and Service Module and blasted off into orbit around the Earth. They had trained to such an extent that they were probably the most well trained of all the astronauts to date, and it showed in their near flawless flight. Their flight evaluated all the major systems with the exception of the Lunar Module, and carried out the first live TV commercial from space. The crew also drank water that had been produced as a by-product of the fuel cells. They carried out optical rendezvous experiments, platform realignment and sextant tracking of another vehicle.

Above left: *Apollo 7* on its way to the launch site on board the Crawler. (NASA)

Above right: *Apollo 7* launching from Cape Canaveral. (NASA)

WALTER CUNNINGHAM
Waltrer Cunningham - Apollo 7.

Above: The crew of the first manned Apollo mission: Don Eisele, Wally Schirra and Walt Cunningham. (NASA)

Left: Photograph of Walt Cunningham signed to the author. (Author)

MANNED AND UNMANNED FLIGHTS TO THE MOON

One of the things that the crew discovered that was to be important for future flights, was the need to exercise and keep muscles in trim. Whilst they slept, they discovered that because of the lack of gravity their bodies invariably went into the foetal position, consequently they woke up with lower back and abdominal pains. The answer was to workout on a stretching device called an Exer-Genie.

The development of the Apollo spacecraft had created more room for the crew to move about in and was in three compartments, the top compartment [forward], the middle and the rear [aft]. The centre compartment was the one occupied by the crew and was a bit larger than the inside of a large car. The sleeping facilities consisted of using sleeping bags slung beneath the outboard couches, but Walt Cunningham decided that he preferred to sleep in his couch using the shoulder and lap restraints. The only problem was, that when two of the crew were sleeping in their couches, they got in the way of any spacecraft operations that needed to be carried out. Although Mission Control suggested that all three astronauts should sleep at the same time, Schirra wanted one of the crew on duty at all times and so they developed their own sleeping rota.

Nearly all the systems in the spacecraft operated exactly as planned, although there were a few annoying problems. One of these were the noisy fans in the environmental section and so first one was switched off and then the other. This caused the coolant lines to 'sweat' forming little pools of water on the deck of the spacecraft. These were

In-flight shot of *Apollo 7* astronauts in their spacecraft. (NASA)

of little concern as this problem was expected after earlier tests had been carried out in the altitude chamber back on earth and the water was quickly vacuumed up using the urine dump hose.

The use of the television camera to show people back on earth their way of life whilst in space was initially felt to be an invasion of their privacy, but after a while they got used to it and would send messages written on cue cards.

Attached to the Command and Service Module [CSM] was the Spacecraft Lunar Module Adapter [SLMA] which housed the LM. Constructed from 1.7 inch [43 mm] thick aluminum, the SLA consisted of four fixed 7 ft [2.1 m] tall panels attached to the instrument unit on top of the Saturn S-IB rocket. This in turn was connected to four hinged 21 foot [6.4 m] panels that opened like flower petals so that the LM could be extracted. Once in orbit the CM pilot would press the CSM/LV button of the instrument panel, which in turn would fire the detonating cord and separate the launch vehicle from the CSM.

As with all previous unmanned Apollo flights, *Apollo 7* did not carry an LM, but it was to carry out a simulated test on opening the SLA adapter and carry out a practice rendezvous with a dummy docking target. It was during this test that a major problem occurred when one of the four hinged panels failed to deploy fully. It was decided to dispose of the hinged panel system and for all further Apollo missions the panels would be fitted with explosive charges that when activated, would send the panels a safe distance from the spacecraft, enabling the extraction of the LM to be carried out safely.

There were also some personal problems that affected the flight, inasmuch as the commander, Walter Schirra, developed a very heavy cold during the flight and was, on occasions, extremely irritable with Mission Control, cutting off transmissions. The other two members of the crew also contracted the cold virus and because of the weightless conditions, suffered a great deal more than had they been on Earth. The mucus, which invariably accompanies a heavy cold, can be easily discharged from the head when on Earth, but when in space it just fills the nasal passages and does not drain from the head. The only relief from this was to blow their noses hard which was extremely painful to the eardrums. This highlighted one of the problems that could be caused by a simple cold on spaceflights, especially if the flights were to be lengthy ones.

With the effects of their colds dragging on, the concern amongst the crew was that during re-entry, the wearing of their helmets would prevent them blowing their noses in an effort to relieve the pressure on their ears and their eardrums might burst under that pressure. Schirra was concerned and told Deke Slayton in Mission Control that they would not be wearing their helmets during re-entry, despite his attempts to dissuade them.

After 164 orbits of the Earth the spacecraft re-entered the Earth's atmosphere on 22 October 1968 without incident, and landed 8 miles [13 km] from the recovery ship, the aircraft carrier *USS Essex* [CV-9].

The *Apollo 7* crew were the first to transmit live television from space and the first to have drinking water produced as a by-product of the spacecraft's fuel cells.

Apollo 7 spacecraft being secured with a flotation collar by divers after splashing down in the Atlantic Ocean. (NASA)

Another space project made the headlines at the beginning of 1969, the Manned Orbiting Laboratory [MOL]. Unlike the other space projects, it made the headlines because its cost was spiralling out of control, so in June 1969, the programme was cancelled. A great deal of time and money had been spent on the MOL project, but in the end it was abandoned mainly because the cost of the Vietnam War was eating into the United States budget and also for lack of progress. The USAF pilots who had been assigned to the project could be re-assigned, if they wished, to the NASA astronaut programme. The Russians however had been looking at the MOL project very carefully and took on board a number of ideas when they created their first space station – *Salyut 1*.

When news reached Moscow about the intended flight of *Apollo 8*, the Russians dismissed it as pure adventurism as they knew that the Americans had no experience of flying unmanned spacecraft around the Moon. As if to emphasise their intentions, the Russians launched *Zond 6* on 10 November 1968 to orbit the Moon and photograph possible landing sites. The Russian press made no secret of the fact these latest flights were rehearsals for a manned orbital flight of the Moon. Problems arose with the telemetry when the high-gain antenna failed to deploy and an alternative had to be found. The orbit of the Moon went without a hitch, but on the return journey back to Earth, another problem manifested itself when the temperature of the hydrogen peroxide propellant dropped from +20 degrees to a dangerously low -5 degrees. The problem was solved by changing the orientation of the spacecraft relative to the Sun. On re-entry the spacecraft suddenly, and inexplicably, depressurised and all contact was lost. It was found two days later and recovered. After a meeting with engineers, technicians and cosmonauts it was decided that because of the problems experienced by *Zond 6*, a manned mission would be put on hold.

The Americans however continued with their Apollo space programme with the launch of *Apollo 8 [AS-503]* on 21 December 1968, with astronauts Frank Borman, James A. Lovell, Jr. and William A. Anders on board. Michael Collins had originally been the CM pilot on the crew, but he had been replaced by Jim Lovell after Collins suffered a cervical disc herniation that required back surgery. This replacement created a situation where for the first time a Gemini crew, *Gemini VII*, of Lovell and Borman, had been assigned to fly together again. This was also the first time that the Saturn V rocket was used to launch a crew into space, *Apollo 7* having been launched using a three-stage Saturn IB. *Apollo 8* was also carrying a non-functional Lunar Module Test Article [LMTA] and although initially there were some thoughts about carrying out a transposition, it was decided that *Apollo 9* would be assigned that task. The *Apollo 8* spacecraft separated on reaching orbit, sending the S-II first stage into the Atlantic, then the second stage was placed into an Earth parking orbit 20 minutes later by the S-IC whilst all the systems were re-checked, then the S-IVB stage was re-ignited and

Apollo 8 atop a Saturn V rocket on the launch pad. (NASA)

MANNED AND UNMANNED FLIGHTS TO THE MOON

Above: Frank Borman, Jim Lovell and Bill Anders about to board the crew bus to take them to the launch pad. (NASA)

Below: Frank Borman, Jim Lovell and Bill Anders in their spacecraft awaiting launch instructions. (NASA)

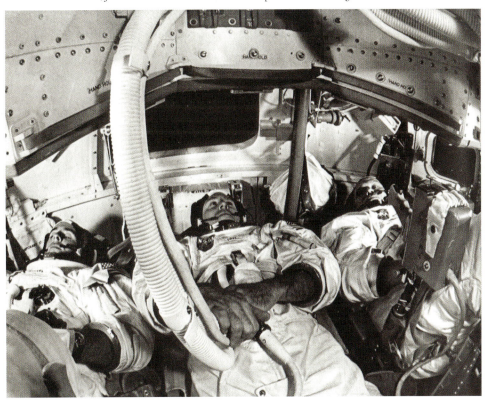

the spacecraft placed into a Lunar trajectory. The crew then became the first men to leave the Earth's gravitational field.

Sixty-nine hours after being launched from Earth, *Apollo 8* approached the Moon and a lunar-orbit-insertion was made to place the spacecraft into a lunar orbit of 193 by 69 miles [310 by 69 km]. As the spacecraft circled the Moon, the crew took numerous photographs of possible landing sites, reference points and landmarks.

On Christmas Eve the spacecraft's communications blacked out as it passed around the dark side of the Moon, the crew becoming the first to actually see the far side. On their way round, the crew took numerous photographs, then, as they came out of the communications blackout, the crew carried out a live television broadcast to Earth. They read the first ten verses of Genesis and at the end wished all the viewers, 'Goodnight, Good luck, a Merry Christmas and God bless all of you – all of you on the good Earth.' It was said later that almost a billion people in sixty-four countries heard the astronauts reading from the bible, either on the radio or on television.

After spending Christmas Day orbiting the Moon the crew fired the CM propulsion system for three minutes and twenty-four seconds, placing the *Apollo 8* spacecraft into an Earth trajectory and increased its velocity to 2,408 miles [3,875 km] per hour. On reaching Earth orbit the Command Module [CM] separated from the Service Module (SM), re-entered the Earth's atmosphere and splashed down in the Pacific Ocean on 27 December, after 147 hours in space. The aircraft carrier *USS Yorktown* [CV-5] reached the recovery area where helicopters from the ship recovered the crew and their spacecraft.

Just over two months later on 3 March 1969, the Americans launched *Apollo 9* [AS-504] with astronauts James B. McDivitt, David Scott and Russell Schweickart

Earth rising above the lunar landscape taken from *Apollo 8* during the first manned mission to the Moon. (NASA)

Apollo 8 re-entering the atmosphere. (NASA)

aboard. The launch had originally been scheduled for 28 February 1969, but was delayed to allow the crew to recover from mild virus respiratory illnesses. The normal launch phase was completed without incident and the S-IVB and Command and Service Module [CSM] *Gumdrop* placed into an Earth parking orbit of 119 by 117 miles [192 by 189 km]. The two spacecraft then separated from the S-IVB and were placed in an Earth-escape trajectory orbit. After all the systems had been re-checked, the CSM separated from the S-IVB, turned around and docked with the Lunar Module [LM] *Spider* which was still attached to the S-IVB. This was the first time the transposition of the Lunar Module [LM] had been carried out. With the separation successfully completed, James McDivitt and Russell Schweickart left the Command Module, entered the Lunar Module through the docking tunnel and carried out tests on the spacecraft's systems. They then carried out a couple of telecasts and fired up its propulsion system. With all the tests completed, the two crew members returned to the CSM. On the following day they returned to the LM and carried out another telecast, before Russell Schweickart carried out an Extra Vehicular Activity [EVA] in which he 'walked' between the two hatches of the spacecraft. During the EVA he took numerous photographs of both the two docked spacecraft, one of which showed David Scott standing in the open hatch of the CSM with the Earth in the background. Schweickart also commented on the rain-squalls that were over the Kennedy Space Center at the time.

Above: Apollo 9 crew of Jim McDivitt, Dave Scott and Rusty Schweickert getting into the crew transfer bus to take them to the launch site. (NASA)

Below: Apollo 9 on the launch pad awaiting a nighttime launch. (NASA)

Right: The Lunar Module in a folded configuration in the third stage rocket waiting to be extracted by the CSM. (Jim McDivitt/NASA)

Below: Rusty Schweickart standing looking out into space at the start of his EVA. (NASA)

Dave Scott standing, looking out at the Earth and space during his EVA. (NASA)

On 7 March, the two astronauts again entered the LM. This time David Scott separated the CSM from the LM and fired the Reaction Control System [RCS] thrusters, placing the two spacecraft 3.4 miles [5.5 km] apart. At one point during the six and half hour separation the two spacecraft were 114 miles [183 km] apart. With all the systems checked out and activities completed the two astronauts docked with the CSM and returned to their spacecraft. The LM was then jettisoned to burn up on re-entry. The remainder of the mission was spent tracking NASA's meteoroid detection satellite *Pegasus III* that had been launched in 1965 and taking multi-spectral photographs of the Earth.

Apollo 9 re-entered the Earth's atmosphere on 13 March and splashed down in the Atlantic east of the Bahamas. The amphibious assault ship *USS Guadalcanal* [LPH-7] recovered the spacecraft and its crew within one hour of their splashdown.

Right: The Lunar Module seen in the distance during tests whilst orbiting the Earth. This was the first time the LM was docked and then undocked with a crew aboard. (NASA)

Below: Apollo 9 about to touchdown in the Atlantic Ocean.

Aboard the *USS Guadalcanal*, NASA carried out a simulation exercise for the recovery of astronauts returning from lunar missions. Three 'astronaut stand-ins', Arthur Lizza, Texas Ward and Paul Kruppenbarcher, together with NASA Flight Surgeon Dr William Carpentier and Project Engineer John Hirasaki spent ten days in the MQF [Mobile Quarantine Facility], a highly modified Airstream trailer. The exercise included the three 'astronauts' climbing into a Command Module mock-up, which was then lowered into the ocean and then recovered as if from a space mission. The three 'astronauts' then put on biological isolation garments and together with the other two NASA employees were placed in the MQF for four days. In the meantime the recovered *Apollo* 9 crew visited them by peering through the observation window in the MQF. After being transferred from the *USS Guadalcanal* to NAS Norfolk and from there to the MSC, Houston, Texas, the five occupants were once again transferred to the Lunar Receiving Laboratory [LRL] where they began a simulated quarantine. The whole exercise was deemed to be a great success and preparations began to prepare for the first of the occupants, which was to be the crew of *Apollo 11*.

The scene was almost set for the attempt to land a man on the Moon, but it was felt that one more mission was required to make sure everything was ready. On 18 May *Apollo 10* (AS-505) was launched, with astronauts Thomas Stafford, Eugene Cernan and John Young aboard. This was the dress rehearsal for what was to be one of the

Apollo 10 crew of Tom Stafford, John Young and Eugene Cernan give a Snoopy mascot a touch on the nose for good luck as they make their way to the crew transport. (NASA)

most dramatic events in world history, when man walked on another world. After going into an Earth parking orbit and the systems re-checked, the S-IVB engine was ignited and the spacecraft placed into a lunar trajectory. One hour later, whilst on course for the moon, the Command and Service Module (CSM) separated from the S-IVB, transposed and docked with the Lunar Module (LM). The S-IVB was then placed into a decaying solar orbit.

After entering a lunar orbit of 68 by 196 miles [110 by 315 km], Stafford and Cernan transferred to the Lunar Module (LM), undocked from the CSM and headed toward the lunar surface. Any thoughts by the two astronauts of continuing their descent to land on the surface of the Moon had been pre-empted by NASA who reduced the amount of fuel aboard the LM so that they could only carry out their planned

Apollo 10 Lunar Module making its way to look at the lunar surface, but that is as close as it got. Had they decided to land there was insufficient fuel in the ascent engine to lift the LM off the surface. (NASA)

Apollo 10 in the Pacific Ocean awaiting recovery by the USS Princeton. (NASA)

assignment. The LM descended to within 8.9 miles [14.3 km] of the surface and then returned to dock with the CSM. The LM was then placed into a solar orbit and the crew prepared to return to Earth. After entering Earth orbit, the SM was jettisoned and the CM re-entered the Earth's atmosphere. Their spacecraft was recovered safely by the *USS Princeton* [CG-59] after splashing down in the Pacific Ocean. The stage was now set for the greatest exploration mission of all time – the first landing on another world by a manned spacecraft from Earth. History was about to be made.

MANNED AND UNMANNED FLIGHTS TO THE MOON

CHAPTER NINE

The First Men on the Moon

There have been a number of stories on how the decision was reached as to who would be the first man to step onto the Moon. The decision was made by the CCB [Configuration Control Board] based on a recommendation by the Flight Crew Operations Directorate.

On 16 July 1969 *Apollo 11* [AS-506] lifted off the pad at Cape Kennedy [the former name of Cape Canaveral], with astronauts Neil Armstrong [Commander], Edwin 'Buzz' Aldrin Jr. [LM pilot] and Michael Collins [CSM pilot] aboard. Four days later Neil Armstrong and Buzz Aldrin transferred to the Lunar Module [*Eagle*] from the CM [*Columbia*] and after checking that all the systems were working properly, they headed toward the Lunar surface and into the history books.

On 19 July *Apollo 11* entered lunar orbit and during the second orbit carried out live colour television broadcasts of the surface of the moon. Whilst this was going on, Aldrin entered the Lunar Module in preparation for the descent to the lunar surface. The following morning Armstrong and Aldrin entered the LM [*Eagle*], checked out all the systems, separated from the CSM and began their descent to the Moon's surface.

Crew of *Apollo 11*: Neil Armstrong, Michael Collins and Buzz Aldrin about to board the crew transport waiting to take them to the launch pad. (NASA)

Sixty miles above the surface of the backside of the Moon, the on-board computers made the first 'burn' to slow the spacecraft down. The 'burn' lasted for 28.5 seconds and the spacecraft coasted towards the front side of the Moon. Inside the LM Armstrong and Aldrin carried out detailed checks of the on-board systems as they rapidly approached the moment when the descent engine would fire up and take them on to the surface. Right on time the descent engine fired, then 26 seconds later the engine went to full power, rapidly de-accelerating the spacecraft. Whilst this was happening, the two astronauts watched the primary and secondary computers. Buzz Aldrin carried out a running commentary to Neil Armstrong, comparing the read-out sequences with their information cards, and all the time the information was being transmitted back to Mission Control.

As they continued their descent, the two astronauts suddenly became aware of a yellow program alarm caution light appearing on the computer. They asked the computer to define the problem and the reply was that the computer was being overloaded with questions and being given too little time to answer them, therefore it could not cope. A second alarm light then appeared and all the time the spacecraft was rapidly descending towards the surface of the Moon. In the spacecraft and back at Houston, everyone waited with bated breath to see what was going to happen next. Back in Mission Control, the man responsible for the computers in the LM, Steve Bales, cut in and told the CapCom, Charlie Duke, to tell the two astronauts to ignore the computers and continue. Two more warning lights suddenly appeared and once again Bales told the crew to override the computers.

At 500 feet [152 m], Neil Armstrong took over manual control and the two astronauts peered out of their observation windows at the rapidly approaching surface. They saw the area they were approaching was strewn with rocks and boulders, but noticed that the area beyond was clear. Neil Armstrong immediately extended the trajectory and touched down in the clear area. Whilst this was happening, Mission Control was monitoring the fuel rate and warned Neil Armstrong that he had only 60 seconds of fuel left. Then came the warning 'thirty seconds', 'ten seconds' and then touch down.

As the spacecraft touched down in the *Sea of Tranquillity*, Buzz Aldrin shut down the descent engine. Then, from Neil Armstrong, the words that the whole world was waiting to hear came over the loudspeakers in Mission Control at Houston, 'Houston, Tranquility Base here. The *Eagle* has landed.'

The relief both in the LM and in Mission Control was immense, and for a few minutes a feeling of complete exhaustion swept through them all. Then the frantic preparations started for the two astronauts to step out on to the surface of the Moon.

It was discovered later that the computer programmers from MIT who designed the on-board computer-landing program in the LM that interrogates the landing radar had never spoken to the computer programmers who designed the rendezvous radar program. This was a major failure on the part of the computer people and one that needed to be addressed and rectified for future missions.

The following day Neil Armstrong stepped out of the LM and stood on the small porch. He gazed at the silent, still panorama around him, then said: 'OK Houston – I'm on the porch.'

Slowly he then climbed down the nine-rung ladder that was attached to one of the landing legs of the Lunar Module and stepped out on to the Moon's surface, the first man to set foot on another world and into the history books, with the now immortal words: 'That's one small step for man – one giant leap for mankind.'

However, the first thing Neil Armstrong did on reaching the surface was to make sure that he could step back up onto the last rung of the ladder. This was because the bottom rung of the ladder was about three feet from the footpad. The shortness of the ladder had been deliberate because in the case of a hard landing, had it been the whole length on the landing leg, it could have been damaged, causing problems for the astronauts trying to step down onto the lunar surface. Twenty minutes later, Buzz Aldrin, after lowering down the Hasselblad camera on a lanyard, stepped down from the ladder to join Neil Armstrong on the Moon's surface. Neil Armstrong, who had clipped the camera to his chest, photographed Buzz Aldrin descending the ladder and stepping on to the lunar surface. The two men gazed around for a few minutes taking in the breathtaking stillness of the lifeless world, summed up in Buzz Aldrin's words 'a magnificent desolation'. They then unveiled a plaque that had been mounted on one of the struts of the Lunar Module that read:

Here men from the planet Earth first set foot on the moon July 1969, A.D. We came in peace for all mankind.

They then raised the American flag [Old Glory], not to claim the territory for their country, but to state that they were the first. Agreement had been reached between nations that in the event of one of them reaching the Moon, no one nation was to claim sovereignty to it. After talking with President Nixon by radiotelephone, the astronauts got to work deploying experimental instruments and collecting rock and dust samples to bring back to Earth.

The two astronauts had quickly adjusted to the one-sixth gravity of the Moon, but found themselves behind schedule. This was because during training the raising of the American flag, the radiophone call with President Nixon and unveiling the plaque, had not been factored into the timescale. It also has to be remembered that all the training was carried out on Earth under almost perfect conditions and now the same activities were being carried out under 'hostile' conditions. After spending twenty-one hours and thirty-six minutes on the surface of the Moon, the two astronauts stored away all the lunar samples and equipment and prepared to blast off and rendezvous with Michael Collins in the CSM *Columbia*. Before leaving the Moon, Neil Armstrong placed on the surface, beside the remaining lower section of the lunar module, two Russian space medals belonging to the deceased cosmonauts Yuri Alexseyevich Gagarin and Vladimir Mikhaylovich Komarov. The medals had been given to American astronaut Frank Borman, Commander of *Apollo 8*, by the cosmonauts'

Buzz Aldrin taking his first tentative steps on the lunar surface. (NASA)

respective wives whilst he was on a goodwill tour of the Soviet Union, requesting that they be placed on the Moon when the Americans eventually reached there.

The ascent stage of the Lunar Module blasted off the Moon's surface and into a lunar orbit to rendezvous with Michael Collins in the CSM. After transferring all the samples and film to the CSM, the Lunar Module [LM] ascent stage was jettisoned to crash back down on the surface. Seismology instruments left on the Moon would register the impact and send the results back to earth.

Placing their spacecraft on a course back to Earth, the crew settled down to a well-earned rest. Their course required only one mid-course correction before they entered earth orbit. The spacecraft splashed down in the Pacific Ocean on 24 July, 15 miles [24 km] from the recovery ship, the *USS Hornet* [CV-12]. Once aboard the aircraft carrier, the three astronauts, together with a doctor and a technician, all

MANNED AND UNMANNED FLIGHTS TO THE MOON

Apollo 11 Lunar Module lifting off the Moon to rendezvous with Michael Collins in the CSM. (NASA)

the lunar rock samples and their EVA suits, were placed inside a Mobile Quarantine Facility [MQF], which was one of four converted Airstream travel trailers that had been created to house astronauts returning from lunar missions. The MQF contained living and sleeping quarters, a kitchen, and a bathroom. By keeping the air pressure inside lower than the pressure outside and by filtering the air vented from the facility, quarantine was assured.

On returning to port in Hawaii, eighty-eight hours later, the *USS Hornet* unloaded the MQF, which was then taken to Hickam Air Force Base where it was loaded onto a Lockheed C-141 Starlifter. It was then flown to Ellingham Air Force Base, Texas and taken by road to the Lunar Receiving Laboratory [LRL] at the MSC, Houston, Texas. The astronauts were then transferred from the MQF, together with the lunar samples that they had brought back with them, into the LRL. The whole purpose of the quarantine was, that in the highly unlikely event that any bacteria that may have been picked up on the Moon was brought back to Earth, it would be contained to see if it was harmful. The five occupants of the LRL stayed in the facility for the next twenty-one days, until doctors were satisfied there were no ill-effects. The quarantine restrictions were dispensed with after the *Apollo 14* mission when extensive tests showed there were no issues.

The lives of the three *Apollo 11* astronauts would never be the same again. The three men had been trained for every aspect and eventuality of the flight to the Moon, but they were not trained for the adulation they would receive when they returned to Earth. Buzz Aldrin even submitted a travel claim for $33.31 to NASA, which was paid.

GRUMMAN AIRCRAFT ENGINEERING CORPORATION
PURCHASE REQUISITION

A374979

NOTE: SEE CORPORATE PROCEDURE M110 BEFORE COMPLETING THIS FORM				
SELLER INVITED TO QUOTE	BUYER	CODE	TELEPHONE - AREA CODE 516	DATE REC'D IN PURCH.
1) North American Rock	North American Rock		LR 5-	
2) Pratt & Whitney	CONTRACT NO. UR B(OO) B(OO)	PRIORITY 1	DATE RELEASED TO TYPE 4/13/70
3) Beech Aircraft	SUBJECT TO GOVERNMENT INSPECTION AT X YOUR PLANT GAEC NONE	TERMS Cash
SELLER AWARDED PURCHASE ORDER	PURCHASE ORDER NUMBER	DATE

SHIP TO Hou - MSC

VIA LM-7, USS Ivo Jima, GOVAIR

SELLER PROMISE Water Again

DELIVERY REQUIRED AT GAEC None

ITEM NO.	QUANTITY	UNIT	PART NO.	DESCRIPTION	I-T	ACCT. NO./JOB NO.	TAX CODE	UNIT PRICE
1	400,001	M1		Towing, $4.00 first mile, $1.00 each additional mile Trouble call, fast service				$400,004.00
2	1	KWH		Battery Charge (road call + $.05 KWH) customer's jump cables				4.05
3	50#	#	BX	Oxygen at $10.00 /lb				500.00
4	1			Sleeping accommodations for 2, no TV, air-conditioned, with radio, modified American plan, with view		NAS-9-1100		Prepaid
5				Additional guest in room at $8.00/night (1) Check out no later than noon Fri 4/17/70 accommodations not guaranteed beyond that time				32.00
6				Water				No Charge
7				Personalized "trip -tik", including all transfers, baggage handling and gratuities				No Charge
				Sub Total				$400,540.05
				20% commercial discount + 2% cash discount (net 30 days)			(-)	83,118.81
				total				$312,421.24
				No taxes applicable (government contract)				

SUGGESTED SOURCES/REMARKS (INCLUDE CWA NO. IF APPLICABLE)

RECEIVING DIVISION USS Ivo Jima

REQUESTED BY NASA (MSC)

APPROVED BY

VIA Air Express

PLT. & DEPT. NO.   EXT.   DATE

DATE

Grumman's invoice for towing Rockwell's CSM - Apollo 13.

Invoice sent to Rockwell by Grumman claiming towing charges for *Apollo 13*. (NASA)

TRAVEL VOUCHER

STANDARD FORM 1012-A

VOUCHER NO. 014501

DEPARTMENT, BUREAU OR ESTABLISHMENT

NASA - Manned Spacecraft Center

PAYEE NAME
Col. Edwin E. Aldrin 00016

MAILING ADDRESS PLEASE MAKE CHECK PAYABLE TO:
Nassau Bay National Bank
P.O. Box 58009
Houston, Texas 77032   Account #1-0346-9

OFFICIAL DUTY STATION
Houston, Texas

APPLICABLE TRAVEL AUTHORIZATION(S)
NO. X-24002   DATE 6/16/69

SCHEDULE OF EXPENSES AND AMOUNTS CLAIMED

DATE	NATURE OF EXPENSE		AMOUNT CLAIMED
	Lv: Residence	0445 POV	
	Ar: SAFB	0500	
	Lv: SAFB	0530 Gov. Air	
	Ar: Cape Kennedy, Fla.	0300 Gov. Spacecraft	
	Lv: Cape Kennedy, Fla.	0832 Gov. Spacecraft	
	Ar: Moon	1325	
	Lv: Moon	2400 Gov. Spacecraft	
	Ar: Pacific Ocean	0600	
	Lv: Pacific Ocean	0800 USN Hornett	
	Ar: Hawaii	0900	
	Lv: Hawaii	1200 USAF Plane	
	Ar: SAFB	0100 Gov. Veh.	
	Lv: SAFB	0215 Gov. Veh.	
	Ar: LRL	0300	

Government meals and quarters furnished for all the above dates.

POV was used for 160 miles official vicinity travel at Cape Kennedy, Fla.

POV authorized for official vicinity travel at Cape Kennedy, Fla. in lieu of rental car.

Thomas P. Stafford

Grand total to face of voucher   33.31

TRANSPORTATION REQUESTS ISSUED

POINTS OF TRAVEL
Houston, Texas   Cape Kennedy, Fla.
Moon
Pacific Ocean
(USN Hornett)
Hawaii
and return to Houston, Texas

AMOUNT CLAIMED   33.31

DIFFERENCES:

NET TO TRAVELER   33.31

APPROVED

AUG 2 6 1969   C. W. Bird

ACCOUNTING CLASSIFICATION
0190-00-00-00-CA-9031-2811

Buzz Aldrin's travel claim for his trip to the Moon – which was paid. (NASA)

When the *Apollo 11* capsule had re-entered the Earth's atmosphere, it generated energy equivalent to approximately 86,000 kilowatt hours of electricity – enough to light the city of Los Angeles for about two minutes; or the energy generated could have lifted the entire population of the United States ten and half inches off the ground. By today's standards, the modern mobile phone is technically more powerful than all the computers that were on board *Apollo 11*, such is the world of technological progress.

In Russia, the news that the Americans had reached the Moon, and two of their astronauts had walked on its surface, was greeted with a mixture of disappointment and, in some respects, relief. The pressure of being the first to get a man to the Moon was now off and they could now concentrate on other aspects of space. There was still the hope that they might send their own cosmonauts to the Moon and with that in mind they launched *Zond 7* on 7 August 1969. This was to be a straight fly-by and although there were a few problems, this was one of the most successful of all the *Zond* mission spacecraft. After orbiting the Moon and taking numerous colour and black and white photographs, the spacecraft returned to Earth and carried out a perfect soft landing in Kazakhstan. During the return flight a problem arose when the temperature in the hydrogen-peroxide tanks started to fall, but by rotating the spacecraft they kept the tanks in direct sunlight. However it damaged the seal around the door to the cabin, resulting in the loss of all the pressure, killing all the insects that were being carried as an experiment. A further problem arose when the parachute ejected prematurely and the spacecraft crashed. Fortunately all the film was retrieved and found to be in perfect condition, so despite the issues the end result had been achieved and so the mission was deemed to be a success.

The next mission to the Moon was on 14 November 1969, when *Apollo 12* [AS-507] blasted off from the pad at Cape Kennedy with its all-Navy crew of Charles 'Pete' Conrad, Jr. [Commander], Alan L. Bean [LM pilot] and Richard F. Gordon [CM pilot], aboard. Thirty-six seconds after liftoff the whole spacecraft experienced a total loss of electrical power after being struck by lightning and again at fifty-two seconds, but power was quickly restored and the spacecraft headed towards an Earth parking orbit.

After carrying out all the necessary re-checking that accompanied every mission, the S-IVB propulsion system was fired up and the spacecraft placed on a Lunar trajectory. One hour into the flight Pete Conrad and Alan Bean transferred to the LM [*Intrepid*] and then proceeded with Richard Gordon, to separate from the S-IVB and transpose with the CSM [*Yankee Clipper*]. Once this had been achieved the S-IVB was sent on its way into a solar orbit, and the CSM with the LM attached went on its way to the Moon. During the transposing the LM crew carried out a live television broadcast showing color pictures of the actual transposing, the interior of the LM, the Earth and the Moon as the two spacecraft raced towards it.

After entering orbit around the Moon, Conrad and Bean undocked the LM [Lunar Module] from the CSM [Command and Service Module] and headed toward the lunar surface. The spacecraft touched down in the *Ocean of Storms* just 590 ft [180 m] from the *Surveyor III* spacecraft that had landed there three years earlier. Conrad was some inches shorter than Neil Armstrong, and as he negotiated the last rung on the LM ladder he exclaimed as he touched the surface:

> Whoopee! Man, that might have been a small one for Neil, but it was a long one for me.

Remembering what Neil Armstrong had done when he first stepped onto the lunar surface, Pete Conrad made sure that he could reach the last rung of the ladder, as did Alan Bean who followed him minutes later. The one-sixth gravity of the Moon helped considerably. The two astronauts carried out two EVAs [Extra Vehicular Activities]

Apollo 12 Lunar Module heading towards the Moon's surface. (NASA)

in which they deployed ALSEP [Apollo Lunar Surface Experiments Package]. Much of the first EVA was spent deploying a set of experiments, some of which continued to radio data back to Earth until September 1977. Among these was a seismometer, which detected thousands of moonquakes and helped to determine the structure of the Moon's interior. Other instruments measured the solar wind and the Moon's tenuous atmosphere. On the second EVA, the crew explored the landing site out to a distance of 1,300 feet [396 m] from the lunar module. They also recovered parts from the *Surveyor III*, including its TV camera and soil scoop. These were taken back to Earth so that they could assess the effects of being on the lunar surface for three years. During the two EVAs Conrad and Bean collected 75 pounds [34 kg] of lunar samples. These rocks were essentially all basalt, a common type of volcanic rock, and were different in composition [less titanium] and younger [3.1–3.3 billion years] than the 3.6–3.9 billion year old *Apollo 11* samples. Above them, orbiting the Moon in the

Alan Bean clambering down the ladder to step out on to the Lunar surface. (NASA)

Alan Bean examining the *Surveyor* spacecraft that had landed some years earlier. (NASA)

Apollo 12 spacecraft, Richard Gordon carried out a lunar multi-spectral photography experiment, photographing future landing sites.

After the first day of exploration and collecting of samples, the two astronauts returned to the LM. The excitement of the day continued whilst in the spacecraft and although they knew that they needed to rest, they found it difficult to settle. They were given permission to remove their helmets and gloves, but it had been decided that they should keep their suits on, because it was feared that the zippers could be clogged with Moon dust and cause a deterioration that prevented them closing securely and giving a good seal.

After thirty-one hours and thirty-one minutes of exploring the lunar surface and collecting samples, the two astronauts lifted off to rendezvous with the CSM. Alan Bean recalled the moment he fired the engine on the ascent stage of the LM:

> I recall looking out of my window, the right-hand one, during lift off and seeing a ring of bright orange, silver and black flashes of light expanding rapidly outwards. It reminded me of the effect I'd seen when I dropped a rock into calm water. It took me a moment to realise that the bright flashes were glints from the pieces of the metal foil insulation blasted from the descent stage by the ascent engine. It was one of the most exciting moments in my space flight.

After rendezvousing with the CSM and transferring the lunar rocks and themselves into the CM, the LM [*Intrepid*] was jettisoned and sent crashing back on to the surface of the Moon, where the reverberations, which lasted thirty minutes, were recorded by a seismometer.

After an uneventful return trip, the crew landed safely in the Pacific Ocean on 24 November and were picked up by the aircraft carrier *USS Hornet* [CV-12]. The MQF-002 was aboard the aircraft carrier when the three astronauts stepped onto the deck and they were immediately ushered into quarantine facility. They were the second crew to go into the quarantine.

Alan Bean was not only an astronaut, a test pilot and aeronautical engineer, but was a highly respected artist. In this painting he shows the moment the ascent stage of the Lunar Module lifted off the Moon. He was the first artist in all of art history to visit another world, return to Earth and paint what he saw and experienced.

The Apollo programme was suddenly subjected to a dramatic turn of events when on 11 April 1970, *Apollo 13* [AS-508] was launched from Cape Kennedy for the third planned landing on the Moon, with astronauts Jim Lovell [Commander], Jack Swigert [CM pilot] and Fred Haise [LM pilot] on board. Jack Swigert, who was the CM pilot, had been part of the back-up crew together with Charlie Duke and John Young. Then Charlie Duke suddenly contracted German Measles and after tests it was discovered that although Jim Lovell and Fred Haise were immune, Ken Mattingly was not and so the decision was made to move Jack Swigert up to replace him. Although

Painting by Alan Bean depicting the ascent stage of Apollo spacecraft lifting off the Lunar surface. (Alan Bean)

Ken Mattingly showed no symptoms, such were the stringent medical and safety requirements, it was decided to err on the side of safety and make the change. The virus never manifested itself with Ken Mattingly, but he later proved to be invaluable to the mission.

During launch an incident occurred that almost resulted in a launch abort. The second stage engine started to oscillate violently, that luckily caused it to shut down early. The engine, which weighed two tons, was bolted to the massive thrust frame, and was bouncing up and down at 6–8g. This caused the frame to flex 3 inches [76 mm] at sixteen Hz! Fortunately, after three seconds of these oscillations the engine's low chamber pressure switch tripped. Fortunate was the right word, as the switch had not

Apollo 13 launching. (NASA)

been designed to trip in this manner – but it did, and the engine automatically shut down. If the oscillations had continued, there is no doubt that it would have torn the Saturn V rocket apart.

With the problem resolved, the launch went ahead without further incident. After entering an Earth parking orbit, all the systems were checked out before the S-IC stage sent the spacecraft on a translunar trajectory. The separation from the S-IC and the transposition of the CSM [*Odyssey*] with the LM [*Aquarius*] went according to plan and was televised. Everything went perfectly until just after 56 hours into the flight, when there was a loud bang. Jack Swigert's apprehensive voice came over the radio, 'OK, Houston, we've had a problem here.'

Because there was no response from mission control, Jim Lovell repeated it: 'Houston, We've had a problem. We've had a main B Bus undervolt.'

It was only after the second call that Houston replied. This was the first indication to Mission Control that there was something seriously wrong. It was discovered later that the Service Module [SM] oxygen tank had ruptured because one of the thermostatic switches in the tank had been welded shut by an electrical arc when 65-volt DC power was applied through a switch designed to take only 28 volts DC. It was then thought that this probably caused the heater tube assembly to overheat, which in turn severely damaged the Teflon insulation on the fan motor wires. This in all probability short-circuited and ignited the insulation and a portion of the tank. The rapid expulsion of high-pressure oxygen that followed would have blown off the outer panel to Bay No.4 of the SM, causing a leak in Oxygen Tank No.2.

There was no way of repairing the damaged tank and no way of turning the spacecraft round and heading back to Earth. The crew had to go on. The Moon landing was now out of the question; it was now a matter of survival.

Fortunately they had extracted the LM *Aquarius* from the S-IC stage and the three-man crew moved into it, which created a problem because it was only ever designed to accommodate two persons. The auxiliary propulsion system on the S-IC was fired up and the remnants of the rocket were put on a collision course with the Moon. Three days later the S-IC crashed into the surface of the Moon at a speed of 5,600 mph [9,012 kph], creating an explosion equivalent to 7.7 tons of TNT. The shock was measured on a seismometer that had been left by *Apollo 12*, which recorded the ensuing vibration lasting for three hours and twenty minutes – a response that baffled seismologists back on earth after they had received the data.

The life support systems in the LM had to be modified to support the three astronauts. The LM was powered up and the CSM powered down. This of course reduced the heating and the astronauts suffered from the cold because of this, but it was necessary to conserve electrical power as it would be needed when they prepared for their re-entry to earth.

The CSM, with the LM attached, headed for the Moon, approaching its western edge. All thoughts of landing on this barren world had been forgotten as Jim Lovell and Jack Swigert carried out their alignment checks by identifying the stars. Fred Haise in the meantime was going through his checklist for the engine burn.

Back in Mission Control, teams of controllers were standing by to relieve the duty controllers, but such was the concern and dedication, that the teams of controllers were reluctant to leave their positions in case something happened during the changeover.

As *Apollo 13* started to go around the dark side of the Moon the crew lost radio communication with Earth, but not before Mission Control had given them the figures to put into the onboard computer for the engine burn – 5 seconds at minimum thrust, 21 seconds at 40 per cent thrust and 4 minutes at maximum thrust.

When the sun disappeared behind the Moon, the blackness emphasised the vast amount of stars that appeared and the three astronauts watched spellbound as the vast curtain of stars unfolded before their eyes. Beneath them the desolate plains, craters and hills passed silently, while the crew took photograph after photograph. Emerging from the dark side of the Moon, Houston called up and established that

The nearest *Apollo 13* got to the Moon. (NASA)

they were ready to carry out a mid-course correction. The engine burn that would send them on their way back to earth would follow this.

One of the major problems that faced the astronauts was the amount of oxygen they had left and the amount of carbon dioxide they were creating. In the CSM there were lithium hydroxide canisters that basically scrubbed the expelled air and removed the carbon dioxide leaving the oxygen. The problem was that the round shaped canisters in the CSM would not fit in to the square shaped receptacles in the LM. The engineers on the ground worked out a system by which the CSM canisters could be connected to the canisters in the LM by means of cardboard, duct tape, hoses and plastic bags. By turning on the fan in the CSM they could pump the carbon dioxide through the canisters – it worked.

Fitting the canisters together in a laboratory was not the same as doing it in a confined space, weightless and in almost freezing temperatures. The whole idea was passed into the simulator where the whole package was put together by Ken Mattingly, in as near similar circumstances as could be managed. When fitted successfully, the information and instructions were passed to the three astronauts and the project completed. With oxygen levels rising and the carbon dioxide levels falling, the crew of *Apollo 13* breathed easier in more ways than one.

Optimism was now rising, but this was short lived when another bang was felt. At first there was no explanation as everything appeared to be normal, but then it was noticed that the amps reading from number two battery was falling rapidly. Fortunately there were three other batteries, and although number two battery had been damaged, it was still functioning – albeit only just.

As the spacecraft approached Earth, the crew powered up the CSM and left the LM, saying goodbye to their 'lifeboat'. Just before re-entry they jettisoned the Service Module [SM] and for the first time, were able to see the extent of the damage that had almost cost them their lives. An entire panel had been blown off, exposing the fuel cells. They took a number of photographs of the SM before re-entering the earth's atmosphere. For a long tense moment, as they came through the communications blackout, families, friends and fellow astronauts held their breath, as it was not known how much damage may have been caused to the spacecraft's heatshield. Then as the parachutes deployed and unfurled, communications were restored and the spacecraft splashed down in the Pacific Ocean. The aircraft carrier *USS Iwo Jima* [CV-46] was standing close by and in less than an hour the three astronauts were aboard ship. The greatest rescue mission in history had been accomplished. There was one sobering thought however, if the explosion had happened *after* the crew had visited the Moon, then there is no way they would have survived, as both sections of their 'lifeboat' would have already been left behind on the Moon's surface!

After a fraught journey, but thanks to the ingenuity of the engineers and the dedication of the teams of personnel back at Houston, the crew returned safely to Earth. One person who should be singled out for praise was fellow astronaut Ken Mattingly, who had originally been a member of the crew, but because he had been in contact with someone who had measles, it had been decided that it was too much of a risk to allow him to go on the mission. When the problem arose, Mattingly placed himself in the

Right: The *Apollo 13* service module showing the damage caused by the explosion. (NASA)

Below: *Apollo 13* crew of Jim Lovell, Fred Haise and Jack Swigert being recovered after their epic rescue mission. (NASA)

The *Apollo 13* spacecraft being recovered by the aircraft carrier *USS Iwo Jima* after its epic rescue. (NASA)

simulator attempting to emulate the same conditions as those being experienced by his fellow astronauts aboard *Apollo 13*, even down to using the identical torch the crew were using. When different ideas were broached to help them survive he carried them out in the Lunar Module simulator to see if they would work. Ken Mattingly spent almost the same length of time in the simulator as did the astronauts throughout their ordeal in space.

It is interesting to note, that despite the continuing 'Cold War' that existed between Russia and America at the time, the Soviet Premier Alexsei Kosygin sent a message to the US Government saying:

> I want to inform you the Soviet Government has given orders to all citizens and members of the armed forces to use all necessary means to render assistance in the rescue of the American (*Apollo 13*) astronauts.

The Russians, who were monitoring the conversations between the MSC and *Apollo 13*, then shut down their transmissions so as to leave the frequency clear of any interference. In addition they immediately offered their help unreservedly, both at sea and in the air, in the event of the spacecraft going off course after re-entry and coming down in their part of the world. They also diverted two cargo ships that were in the South Pacific at the time, into the expected landing area to help if required. The feeling between the astronaut and cosmonaut fraternities was that this crisis transcended politics and had become a humanitarian one. A number of other countries also offered their help by dispatching naval vessels into different oceans just in case. Fortunately their assistance was never required, but it showed that when a crisis arose most countries would put aside their differences.

Later the Grumman Company, who had made the Lunar Module 'lifeboat', sent Rockwell, who had made the Command and Service Modules, a tongue-in-cheek invoice for towing the damaged modules back to Earth. At the bottom of the invoice it was noted: 'Lunar Module checkout no later than noon Friday. Accommodations not guaranteed beyond that time'.

Then the public relations director of its Downey, California space division, Earl Blount, issued an equally tongue-in-cheek statement saying that Grumman, before sending such an invoice, should remember that North American Rockwell had not yet received payment for ferrying LMs on previous trips to the Moon.

The near loss of the *Apollo 13* mission had been turned into a magnificent success for mission control, bringing together people from all sections of the space industry in one common cause – the saving of lives.

The Russians by now had abandoned their manned lunar programme and concentrated instead on sending unmanned spacecraft to the Moon.

On 12 September 1969, they launched *Luna 16*. This was one of the most flawless missions of the Luna programme. After orbiting the Moon the spacecraft landed in the *Mare Fecunditalis* region. This was the first robotic landing by the Russians that successfully collected and returned a sample of lunar soil to Earth. This was followed on 10 November 1970 when the Russian spacecraft *Luna 17* soft landed on the surface of the Moon in the *Mare Imbrium* region and released a lunar rover: *Lunokhod 1*. Powered by solar power, the unmanned rover was to tour the lunar surface for the next eleven days sending back photographs. By the time the mission ended, *Lunokhod 1* had travelled roughly 6.5 miles [10.54 km], sending back 20,000 TV pictures and 200 TV panoramas, and had taken more than 500 lunar soil tests. Two years later, on 16 January 1973, the programme was repeated when *Lunokhod 2* landed on the surface of the Moon and carried out similar experiments and investigations as its predecessor. During its time on the surface it covered approximately 22.9 miles [37 km] on the lunar surface.

The launch on 31 January 1971 of *Apollo 14* (AS-509) with astronauts Alan B. Shepard, Jr. [Commander], Stuart Roosa [CM pilot] and Edgar D. Mitchell [LM pilot] aboard, was to put the Americans back in the driving seat as far as manned spaceflight was concerned. This was the most inexperienced crew to go into space as only Alan

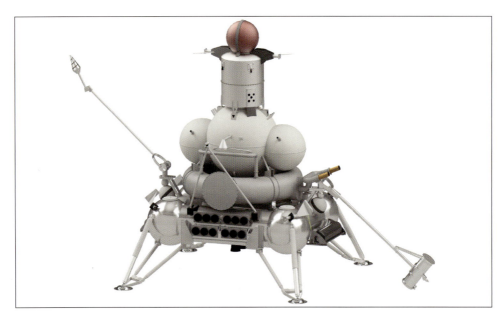

The Russian spacecraft *Luna 16*. (NASA)

Lunokhod, the first Russian unmanned lunar rover on the Moon. (NASA)

Shepard had been in space before and that was only for 15 minutes. As it turned out it was to be one of the most successful and almost trouble free Apollo missions.

On reaching Earth parking orbit, as with all previous missions, the astronauts checked out the systems before firing up the S-IVB and putting the spacecraft into a translunar trajectory.

During the transposition of the LM [*Antares*] and the CSM [*Kitty Hawk*], the crew experienced difficulty in docking. By the fifth attempt, consideration was being given to a possible abort, but after the sixth attempt the latches activated and the two spacecraft were mated. An inspection of the latching mechanism was carried out, but nothing could be found and the cause of the difficulties could not be determined. On entering lunar orbit on 5 February 1971, Shepard and Mitchell entered the Lunar Module and headed down to the lunar surface. Despite problems with a potential short circuit in an abort switch on the Lunar Module *Antares*, and with the landing radar coming on very late in the landing sequence, Alan Shepard and Edgar Mitchell landed less than 98 ft [30 m] from the target point, near Cone Crater in the *Imbrium Basin*. This had been formed when a pro-planet from the asteroid belt hit the Moon some twelve million years ago and landed in the *Fra Mauro* region of the Moon.

After landing there was a forty-nine minute delay because of a communication problem in one of the Portable Life Support Systems [PLSS]. With this problem resolved, the two astronauts carried out the first of their two scheduled EVAs and deployed a TV S-band antenna, a solar wind experiment and of course the American flag. They then moved their next experiment, the lunar surface

Apollo 14 crew: Ed Mitchell, Alan Shepard, Stuart Roosa. (NASA)

Mobile Equipment Transporter [MET] being loaded by Alan Shepard. (NASA)

experiment, 500 ft [152 m] away from the LM. Among the other experiments was a seismometer, which detected thousands of moonquakes and helped to determine the structure of the Moon's interior. Other instruments measuring the composition of the solar wind and the Moon's tenuous atmosphere and plasma environment were set up.

The second EVA was carried out the following day and this time they used the Mobile Equipment Transporter [MET] to carry their tools and photographic equipment, and carried out a geological survey of the rim of the Cone Crater. Most of the rocks collected were breccias and impact melted rocks that were formed during the extreme temperature and pressure of the Imbrium impact event and other crater-forming events. Studies of these rocks indicate that the *Imbrium* impact occurred approximately 3.93 billion years ago. A few of the rocks found at *Apollo 14* are basalts. They are volcanic rocks and are generally similar to basalts found at most of the other Apollo landing sites, although the *Apollo 14* basalts were older than those from other sites. Some of the *Apollo 14* basalts have ages of 4.0 to 4.3 billion years. This was a 3 km round trip and on their way back to the LM they re-adjusted the ALSEP.

Just before leaving the Moon, Alan Shepard dropped a golf ball onto the surface and on the third swing, drove the ball a distance of 1,200 ft [366 m]. Having established membership of the most exclusive golf club in the universe, the two astronauts returned

Apollo 14 crew being removed from their spacecraft by Navy divers prior to being collected by helicopters. (NASA)

to the LM. After stowing away 95 lb [43 kg] of lunar rock samples and dust, they lifted off the following day, rendezvoused with the CSM and set a course for Earth.

After transferring all the samples and equipment from the LM to the CSM, the ascent stage of the LM was jettisoned and sent crashing back into the Moon's surface, where the impact was recorded by the ALSEPs left by *Apollos 12* and *14*.

The return to earth was uneventful and a number of experiments and tests were carried out during the flight back. As they approached the earth, the CM and SM separated, leaving the CM to head back to Earth. After spending nine days in space, the spacecraft splashed down just 4.3 miles [7 km] from the recovery ship, the amphibious assault ship *USS New Orleans* [LPH-11].

With the Americans now well ahead in the space race, the Russians concentrated on how to work in space, and developed the first space station. On 19 April 1971 the Russians launched *Salyut 1*. Four days later the first cosmonauts to go to the space station were launched aboard *Soyuz 10*, but because of a problem with the docking mechanism were unable to dock so had to return to Earth. With the docking mechanism modified, the next crew of Valeri Kubasov, Alexi Leonov and Pyotr Kolodin prepared to take *Soyuz 11* to the space station, but during a pre-flight medical Kubasov was found to have a dark spot on his chest X-ray and so with that the whole crew was changed to the back-up crew. The new crew, Georgi T. Dobrovolski, Viktor I. Patsayev and Vladislav N. Volkov, were launched on 6 June, but once again there were problems with the docking mechanism and it took three hours and 39 minutes for docking to be established. After spending twenty-three days aboard *Salyut 1* carrying out a series of experiments, which according to the Russians, were extremely productive, the three cosmonauts closed the hatch of *Salyut 1*. They were not wearing pressure suits when they climbed into their *Soyuz* spacecraft for the journey back to Earth. After completing three orbits of the Earth, *Soyuz 11* entered the atmosphere. Just thirty minutes from landing, whilst still in space, tragedy struck. There had been some concerns regarding radio contact with the crew as they had heard nothing since just before re-entry. On landing in Kazakhstan the recovery helicopters and crews were quickly on the scene and, on opening the capsule, discovered the lifeless bodies of the three cosmonauts still strapped in their seats. It was later discovered that a pressure equalisation valve, which normally opened once the parachutes had been deployed, had been jarred loose whilst in the vacuum of space, releasing all the air in the capsule in less than a minute. There was evidence that the crew had attempted to manually close the valve, but that would have taken several minutes and by then they would have probably all been unconscious. Despite frantic medical attention it was too late and was a bitter blow to the Russians, and indeed the whole of the space fraternity, and once again highlighted the dangers of space exploration. Alexi Leonov and Valeri Kubasov were later to fly on the Apollo/Soyuz Test Project [ASTP] mission.

America continued to carry out missions to the Moon, this time aboard *Apollo 15* [AS-51]. The spacecraft was launched on 26 July 1971, with astronauts David R. Scott [Commander], James B. Irwin [LM pilot] and Alfred M. Worden [CM pilot] aboard. As with all previous missions, the spacecraft went into an Earth parking orbit just twelve

Emergency medical treatment being carried out on the bodies of the *Soyuz 11* crew after they had been discovered lifeless in their spacecraft upon landing. (Gromov Institute)

minutes after lift-off. When all the checks had been completed, the S-IVB propulsion system was fired up and the spacecraft sent into a translunar trajectory.

During the flight to the Moon, the transposition of the LM [*Falcon*] from the S-IC and the mating with the CSM [*Endeavour*] took place. Dave Scott and Jim Irwin entered the LM after entering lunar orbit and prepared to undock from the CSM. Firing up the LM the two astronauts closed the hatch and initiated the undocking sequence, but nothing happened. Puzzled, the two astronauts went through the sequence once again, but still nothing. Al Worden in the CSM checked all the umbilical cords and discovered one had not seated properly. With that resolved the two spacecraft separated and the LM headed for the surface of the Moon. The journey to the surface was controlled by the onboard computer and all seemed well as they waited for the probe on the landing leg to touch the surface, which would then automatically switch on the cabin lights, telling the crew to switch off the

Apollo 15 crew of Dave Scott, Jim Irwin and Alfred Worden in their spacecraft about to have the hatches closed on them. (NASA)

Dave Scott with the Lunar Rover. This was the first time the Rover had been used. (NASA)

descent engine. When the lights came on, Jim Irwin called 'Contact' and Dave Scott pressed the button to switch off the engine. Clouds of dust obscured their view as the LM touched down hard, causing the spacecraft to roll and pitch and the two astronauts to almost fall down. The impact was much harder than was expected and the LM had developed a lean. The two astronauts held their breath as they waited for mission control to check their systems and give the order to 'Stay'. The spacecraft had landed right on the edge of a small crater in the *Hadley-Apennine* region. This was a tense moment because if the lean exceeded forty-five degrees, they would have to abort. They knew that if the LM fell on its side, there was no way they would be able to right it and so breathed a great sigh of relief when the order to 'Stay' came. The two astronauts were to carry out five EVAs and, after unpacking the Lunar Rover Vehicle [LRV], carried out the first real exploration of the lunar surface. There was difficulty in unloading the LRV at first, mainly because of the LM's angle after landing, but after a bit of maneuvering and manipulation, it was placed on the surface. Then it was discovered that the front-wheel steering would not work, but fortunately the rear-wheel steering did, so the two astronauts were able to move. Although Jim Irwin and Dave Scott had trained extensively on a model of the LRV back on Earth, they discovered that this was a whole new ball game and it took a while to acclimatise to both the environment and the vehicle itself. During the first EVA, Jim Irwin commented that on Earth he weighed 160 lb [72.6 kg], his space suit

50 lb [23 kg] and the backpack, including the PLSS [Portable Life Support System], 80 lb [36 kg] and was difficult to walk in, but on the Moon, with its one sixth gravity, there was no problem.

The LRV gave the appearance of a chassis with a wheel in each corner and two seats perched at the rear. The tires were made out of piano wire that had been stretched to form a surface similar to that of a rubber tire. It was powered by sealed electric motors in the wheel hubs and ran on two 36-volt electric batteries. Mounted on the buggy was a complete communications package that kept the astronauts in radio and TV contact with Mission Control. The whole LRV weighed a staggering 455 pounds [206 kg] on Earth, but on the Moon weighed only 76 pounds [34.5 kg], and the cost and development of this little four-wheel drive off-road dune buggy was a mere $8 million.

The LRV could carry two and half times its weight at a speed of 10 mph [16 kph]. It was a weird, flimsy-looking machine but it worked, and it enabled the astronauts to explore farther away from their spacecraft and obtain samples of rock from various locations. One of those, named the Genesis Rock, was age-dated at 4.15 billion years old, plus or minus 25 million years, by the University of New York. The oldest thing ever found on Earth was dated at 3.3 billion years, so another piece of puzzle to the creation of the universe was found.

The LM of *Apollo 15* was different to the previous four models inasmuch as it was designed to spend longer on the Moon. Previous missions had been scheduled for a one day stay, but *Falcon* was kitted out for a three day stay. In addition the astronauts' PLSS had been redesigned for them to spend seven hours on the surface, so knowing all this, Dave Scott set their EVAs accordingly without going onto the lunar surface within a very short time of landing, unlike previous missions.

Over the next few days the two astronauts roamed the lunar surface in their LRV collecting samples from different locations. After the last EVA, and after unloading all the collected samples into the LM, the LRV was stripped of everything that could be taken back to earth, with the exception of the TV camera. That was pointed at the Lunar Module and recorded the ascent stage blasting off the surface of the Moon. Just before the two astronauts entered the Lunar Module for the last time, they placed a plaque on the surface of the Moon inscribed with the names of all the astronauts, both American and Russian, who had died in the pursuit of space exploration. Jim Irwin, who was a deeply religious man, asked Mission Control for permission to hold a religious service on the Moon. They refused saying that there was far too much to be done to allow time for a religious service. After blasting off the Moon's surface, the LM went into orbit so that they could rendezvous with Al Worden in the CSM. As the CSM was about to leave lunar orbit, the crew launched a sub-satellite to carry out scientific studies of the Moon from a low orbit.

After jettisoning the LM, which was sent crashing back onto the lunar surface, the CM returned safely to Earth on 7 August 1971 and was recovered by the amphibious assault ship *USS Okinawa* [LPH-3]. Another successful trip to the Moon had been accomplished. With the success of the *Apollo 15* mission ringing in their ears, the crew were assigned as back-up crew for *Apollo 17*.

Then came the scandal concerning a number of unauthorised commemorative postal covers that were flown to the Moon aboard the LM and the CSM, although there were a number of authorised covers taken. A number of the covers, which were carried in the astronauts' Personal Preference Kits [PPK], had their stamps cancelled whilst on the surface, endorsing the fact they had been on the Moon. There were however a large number [400] of unauthorised postal covers taken and, including the authorised ones, they totalled 650. The whole incident came to light when a collector of space memorabilia wrote to NASA asking if the covers were genuine. It then transpired that a number of astronauts had had dealings with a German by the name of Horst Eiermann, a representative for a company who made space products and was well known to many of the astronauts. He had put them in touch with another German who was a philatelic dealer in Germany by the name of Hermann Sieger. A number of other astronauts had agreed to sign a large number of commemorative covers and blocks of stamps for $2,500.

The *Apollo 15* astronauts then got involved and purchased stamps depicting the space programme and, with the help of their secretaries, fixed them to the illustrated envelopes. Once the stamps had been cancelled in the KSC Post Office, they were signed by the three astronauts and vacuum packed by the Flight Crew Support Team ready to be placed aboard the CM. At the end of the mission, 100 of the signed and endorsed covers were given to Horst Eirmann who in turn sent them on to Sieger. The three astronauts had stressed to Eirmann at the time, that the covers were not to be sold until the Apollo programme was over. A couple of weeks later, the *Apollo 15* crew received bank books from Hermann Sieger containing large sums of money. Unfortunately Sieger, on receiving the covers, promptly sold them to customers on his mailing list. On hearing that the covers were now on sale in Europe, the astronauts promptly returned the money, saying that the terms of agreement had been broken and they wanted nothing more to do with the selling of the covers. It was then that the problems started. There does not appear to have been any effort by Horst Eirmann to ensure that the conditions with regard to selling the 'flown' covers were adhered to by Hermann Sieger. One would have thought that knowing NASA's non-commercial policy, the astronauts would have been well aware of pitfalls in embarking on such a venture, especially with someone in a foreign country. But the damage had been done.

An investigation was started by NASA and a number of the astronauts were asked to return any unauthorised items or postal covers that they had been involved with, with the result that fifteen astronauts were suspended pending further investigations. After an intense investigation by NASA, the three *Apollo 15* astronauts were reprimanded and told that they would never be part of a flight crew again. Al Worden went to work at NASA's Ames Research Center, Dave Scott became a director of NASA's Dryden Research Center and Jim Irwin stayed as a member of the *Apollo 17* back-up crew. He was back-up to Harrison Schmidt, but he knew that he

MANNED AND UNMANNED FLIGHTS TO THE MOON

would never fly again as NASA were desperate to have a scientist explore the Moon and would even reschedule if Schmidt was taken ill. Jim Irwin took the decision to retire from NASA and the Air Force and founded an evangelical organisation called 'High Flight'.

In the meantime the Russians were still photographing and mapping the Moon and on 2 September 1971, they launched *Luna 18* from the Baikonur Cosmodrome. Five days later it went into a lunar orbit and fired its retro-rockets, but they shut down fifteen seconds early, putting the spacecraft into an incorrect orbit. Then followed a fault in one of the attitude control engines causing the spacecraft's orbit to change once again. Despite attempts to soft-land the spacecraft, the failure of another attitude control engine sent it crashing to the surface and was destroyed. Not to be deterred, the Russians launched *Luna 19* on 28 September 1971. This time the mission was to orbit the Moon and study the lunar radiation environment, solar wind and the magnetic field, amongst other experiments. It also sent back photographs and a limited amount of television coverage. A problem arose when a faulty gyro placed the spacecraft into a higher orbit, and despite attempts to rectify the problem, communication was lost and it is assumed it later went into a decaying orbit and crashed.

The launch of unmanned *Luna 20* on 14 February 1972 restored Russia's confidence when the spacecraft went into orbit around the Moon after a flawless flight. The descent stage was equipped with telecommunications equipment, radiation and temperature monitors, a television camera and an extendable arm with a drilling rig and lunar sample scoop attached. After recovering 2 ounces [55 grams] of lunar soil, *Luna 20* blasted off the lunar surface and returned safely to Earth.

America's sixth manned mission to the Moon, *Apollo 16* (AS-511), was launched on 16 April 1972, with the crew of astronauts John W. Young [Commander], Thomas L. Mattingly II [CM pilot] and Charles M. Duke, Jr. [LM pilot]. On reaching Earth orbit the crew put the spacecraft into a holding orbit while they checked out the systems. With all systems checked the S-IVB was fired up and the spacecraft were put into a translunar trajectory. About one hour into the flight the CSM [*Casper*] jettisoned the S-IVB and transposed with the LM [*Orion*]. On entering lunar orbit John Young and Charles Duke entered the Lunar Module and powered up all the systems. On separation the two spacecraft stayed on station whilst the crew evaluated the service propulsion system. There had been a concern about unexpected oscillations in the control system for the main engine, but it was decided that it was safe and could be used if required.

After orbiting the Moon, John Young and Charles Duke set the LM down in the *Descartes* region which had been chosen because it was thought there were signs of there having been volcano activity, which later proved to be unfounded. The landing site had also been selected from photographs taken from *Apollo 14*. There was a delay of six hours before the decision was taken to leave the LM, because of the concerns about the main engine. Mission Control made the decision to go, allowing the two astronauts to take a rest break before carrying out their first EVA. The following

Apollo 16 launching. (NASA)

morning John Young and Charles Duke unpacked the LRV and prepared to carry out an exploration in the region of Survey Ridge, Stone Mountain and North Ray Crater areas. The two astronauts collected a number of samples of rock and it was soon realised that there was no volcanic material in that area. On the second EVA, which lasted seven hours and twenty-three minutes, the two astronauts drove to Stone Mountain, which was thought to have been a volcano, and climbed to 525 ft [160 m]. It had been intended to collect samples but it proved not to have been a volcano, so instead they headed in the LVR to the two million year old, 2,231 ft [680 m] wide South Ray Crater.

The third EVA had to be reduced to five hours and 40 minutes because of the delay in landing. The two astronauts headed for the North Ray Crater which was 3,280 ft [1 km] across and 253 ft [77 m] deep. The impact which caused the crater would have thrown boulders out from deep below the Moon's surface and so a number of samples were sought out and collected. During the three EVA the two astronauts collected 211 lb [96 kg] of Lunar rocks and dust. At one point Charles Duke was heard to say as he looked up at the Earth, 'Man we are a long way from home.'

The ASLEP being deployed with the Lunar Module in the far distance. (NASA)

As with all Apollo missions after *Apollo 11*, one of the experimental packages was the Apollo Lunar Surface Experiment Package [ALSEP]. This consisted of a Passive Seismic Experiment [PSE], Active Seismic Experiment [ASE] and a Lunar Heat Flow Experiment [HFE]. Two of the experimental packages unique to *Apollo 16* were the Traverse Gravimeter Experiment [TGE] and the Surface Electrical Properties [SEP]. The Gravimeter had been developed to study the Earth's internal structure, and the object of this experiment was to study the Moon's internal structure. It was to measure the relative gravity at the landing site and at various other locations. The TGE was mounted on the LRV and as the two astronauts moved around they recorded measurements. A total of twenty-six measurements were taken and the results produced some very productive results.

The SEP experiment consisted of two components, a transmitter which was deployed by the LM and a receiver that was mounted on the LRV. As the LRV

The Lunar Rover being driven along a ridge in the far distance. This was a good example of how the astronauts managed to cover vast distances on the lunar surface. (NASA)

moved around different sites, the transmitter sent electrical signals through the ground which were received by the LRV. The results obtained showed that there was no water in the area of the Moon in which *Apollo 16* landed, to a depth of 1.2 miles [2km], which is not to say that there was no water elsewhere. Holes were also drilled into the surface to a depth of 8 ft [2.4 m] to collect core samples. This was because over millions of years the surface had been covered or mixed with micrometeorites and what scientists and geologists wanted to know what was beneath.

The LRV was used extensively and over a total drive time in the three EVAs of four hours and twenty-six minutes, they covered a total of 22 miles [35.7 kilometres]. Like the previous two LRVs it was parked up and left on the Moon with the remainder of the experiments.

They also transmitted live colour TV pictures back to Earth and, when they departed, their spectacular launch from the surface of the Moon was captured live on TV from the camera mounted on the LRV. After spending seventy-one hours on the Moon and carrying out three EVAs over a period of twenty hours, all using the

LRV, during which the two astronauts carried out a number of investigations and experiments, they blasted off the lunar surface.

After transferring to the CSM the LM was jettisoned, and like the other LMs before it, meant to impact on the Moon's surface. This was so that seismic measurements could take place, enabling scientists back on Earth to try and measure the depth of the Moon's crust. Unfortunately the LM went into a decaying lunar orbit around the Moon. As the *Apollo 16* spacecraft left lunar orbit it deployed a scientific sub-satellite, but that too ran into problems and like the LM went into a decaying orbit. The concerns about the main engine caused the mission to be cut short, but as it turned out there were no problems, and after docking with the CSM, the two spacecraft headed for home. On the flight back, Ken Mattingly made a one hour and twenty-three minute spacewalk where he recovered the film from a mapping camera situated in the CSM. The *Apollo 16* spacecraft returned to Earth safely and splashed down in the Pacific Ocean on 27 April and was recovered by the aircraft carrier *USS Ticonderoga* [CV-14].

Whilst in orbit around the Moon, Ken Mattingly in the CM became aware of flashes of light coming from the Moon's surface, but was unable to pinpoint them. It was never discovered what they were, but suffice to say the UFO theory once again raised its head.

The final manned Apollo lunar explorer *Apollo 17* [AS-512] was launched on 7 December 1972 at 00.33 hours EDT from the Kennedy Space Center. The crew, Eugene Cernan [Commander], Ronald Evans [CM pilot] and Harrison Schmidt [LM pilot], had to spend an additional three hours in the CSM because of a countdown sequence failure. This was the only hardware failure during the entire Apollo programme to cause a launch delay. The launch was a dramatic nighttime launch which lit the skies around the Kennedy Space Center. Earth orbit was achieved and the insertion into a lunar trajectory carried out without a hitch. The transposition of the CSM [*America*] with the LM [*Challenger*] was completed during the trans-lunar flight and again was incident free. A number of scientific experiments were carried out during this part of the flight to the Moon as on all previous missions. *Apollo 17* entered lunar orbit on 11 December and after completing the separation of the LM from the CSM, Cernan and Schmidt made their descent to the surface. The landing site chosen was in the *Taurus-Littrow* area, a site observed and photographed by Al Worden on *Apollo 15*. Harrison Schmidt was the first geologist-scientist-astronaut to go to the Moon. The first EVA began four hours later and, after off-loading the LRV and experimental packages, the two astronauts decided to go ahead with some of their tasks rather than wait until the following day.

The area in which the LM had landed was known as *Taurus-Littrow* and was near the coast of the great frozen sea of basalt, the *Mare Serenitatis*. The valley in which they had landed was deeper than the Grand Canyon back on Earth. The unique visual beauty of this valley was the epitome of an ethereal vision of sheer desolation. Of all the landing sites chosen, it would have been hard to find one that ended the exploration of the Moon in such a memorable way. The LRV was used extensively on all the lunar trips and in one incident when Eugene Cernan dropped a hammer on the fender of the rover and broke it, repairs had to be made. The astronauts pieced together a replacement using duct tape and four stiff maps, then together with the aid of two clamps, they created a makeshift flap, which they then attached to the fender. On hearing of the lunar body repair story, the Auto Body Association of America bestowed a lifetime membership on the two astronauts for their emergency repair – the first and only time a vehicle had been repaired whilst on another world.

The second EVA lasted 7 hours and 37 minutes and it was during this period that the now-famous orange soil was discovered. It was to be the subject of geological discussion for many years to come.

The third and final EVA was on 13 December, during which a great variety of geological samples were taken. One of Eugene Cernan's last acts was to scrawl his daughter's initials in the lunar dust. Just prior to entering the LM for the last time, a plaque was unveiled on the landing gear. It said quite simply:

> Here man completed his first explorations
> of the Moon, December 1972 AD
> May the spirit of peace in which we came
> be reflected in the lives of all mankind.

It was signed by the crew of *Apollo 17* and by the then President of the United States of America, Richard M. Nixon. This last mission, according to NASA, was the longest and most successful of all the manned lunar missions. The return journey to Earth was uneventful and the spacecraft splashed down in the Pacific Ocean close to the recovery ship, the *USS Ticonderoga* [CVS-14].

Back on Earth the American public, although retaining some interest in the on-going space programme, had become apathetic and this was reflected by the three main television networks, who broadcast the *Apollo 15, 16* and *17* moonwalks late at night. ABC television squeezed one of the moonwalks into the halftime of *Monday Night Football*. Some of the new and upcoming television companies continued to broadcast the latter space missions, but in the main interest was waning. The world of men floating in space and walking on the Moon had become commonplace and was no longer the main topic of conversation. It had become normal and seemingly an everyday occurrence and no longer held the majority of the American people in awe. It was decide to cancel *Apollo 18* and enter a different aspect of space – an orbiting laboratory.

On 8 January 1973, the Russians sent their second lunar rover, *Lunokhod 2*, to the Moon aboard *Luna 21*. All went well, with the lunar rover sending back over 80,000 TV pictures and eighty-six panoramic photographs, until it drove into a crater and covered all its solar panels and radiators with fine lunar dust. Despite desperate attempts to recover the lunar rover, all were unsuccessful and the mission was abandoned.

Right: *Apollo 17*'s lunar rover being readied for a trip. (NASA)

Below: *Apollo 17* awaiting recovery with the *USS Ticonderoga* in the background. (NASA)

CHAPTER TEN

Skylab the Space Laboratory

Skylab 1

With the cancellation of *Apollo 18*, thoughts turned to placing a laboratory in orbit around the Earth. Despite the problems that the Russians had experienced with their *Salyut* space station, the American Skylab project, which had been born during the Apollo missions, came to fruition. The unmanned space laboratory *Skylab 1* was launched on 14 May 1973 and, complete with all necessary stores and equipment, went into orbit around the Earth, waiting for its first occupants. It was much roomier than the Russian *Salyut* space station and so was able to contain and conduct many more experiments. It was originally thought that once in orbit it would be regularly visited by astronaut/scientists, but as it turned out, only three crews would ever visit *Skylab*. *Skylab* was exactly what it said it was, a research laboratory in the sky. It was built from the Saturn S-IVB stage of the Saturn V rocket booster that had been left over from the Apollo programme. It consisted of a two-storey accommodation 48 ft long by 21 ft in diameter. The upper section was the laboratory/workshop, whilst the lower section contained the living quarters for three astronauts.

Within one minute of leaving the pad, Flight Controllers of the Saturn V reported 'a strange lateral acceleration' at approximately 63 seconds after lift-off. Within minutes an endless stream of telemetry confirmed there was a problem. A metal shield only 0.025 in [0.6 mm] thick fitted around the orbital workshop area had torn loose. This shield was a critical part of the workshop's protection inasmuch as it was there to protect the spacecraft from small meteoroids and also to protect the workshop from the searing heat of the sun. The workshop had two large solar panels attached to the outside which, during launch, were folded against the sides of the workshop, then once in orbit were extended. Problems arose when radar trackers in Australia reported that the two solar panels of the workshop had not deployed. This meant that there was no power going to the space station.

Temperatures within the spacecraft started to soar to 190° as the sun beat down on the unprotected spacecraft. Controllers on the ground issued a command to *Skylab* to start rotating slowly so as to dissipate the heat, and engineers on the ground devised a kind of parasol that could be fitted over the top of the workshop area.

Attached to the forward end of the workshop was the airlock module. This enabled the astronauts to enter and leave the workshop area without having to depressurise the area. Also in the airlock module were the temperature controls,

Above left: Skylab on the launch pad at Cape Kennedy. (NASA)

Above right: Skylab in orbit. (NASA)

air purification systems, electrical controls and hazard warning systems for *Skylab*. Attached to the aft end of the workshop were the multiple docking modules that enabled Apollo spacecraft to dock at either of two docking ports, one on the end and one on the side.

The living quarters consisted of three bedrooms, a kitchen and a bath. The crew had a veritable warehouse full of clothes consisting of sixty pairs of jackets, trousers and shorts; fifteen pairs of boots and 210 pairs of underpants. The bathroom contained over fifty bars of soap, fifty towels and 1,800 urine and fecal bags. The latter were used to carry out investigations on the bodily wastes of the astronauts for medical and biological reasons. In fact the size of *Skylab* was almost the same as that of a small, two bedroom house.

Skylab II

The first crew for *Skylab* lifted off from Cape Canaveral on 25 May 1973, but right from the start there were problems. When the first astronauts, Captain Charles

'Pete' Conrad [Commander], Commander Paul J. Weitz and Commander Dr. Joseph P. Kerwin, all from the United States Navy, arrived on board *Skylab I*, they found significant external damage. Fitting the parasol was their first task, so donning their EVA suits, Pete Conrad and Joseph Kerwin climbed outside the *Skylab*, fitted the parasol and then released the solar panels that immediately gave power to the space station. *Skylab* was operational and proved conclusively that no matter how sophisticated technology became, when major problems arose it was left to the ingenuity of man to fix them. With the *Skylab* space station stabilised and operational, the first mission was to last twenty-eight days, in which a number of experiments were to be carried out. These included the study of Alpha emissions from the sun during solar flares; UV and X-ray photographs of the sun and X-ray emissions of the sun's lower corona. But it was soon realised that the number of experiments would have to be cut because of the maintenance requirements of *Skylab* itself. The repairs carried out initially would have to be continually monitored. The crew returned to Earth in their Apollo spacecraft having made the *Skylab* ready for the next crew.

Pete Conrad in the orbital workshop during *Skylab 1* mission. (NASA)

Whilst interest was focused on *Skylab*, the launch on 10 June 1973 of *Explorer 49* seemed to go unnoticed. This was to be the final mission to the Moon for the time being and, because of its extremely large antenna, this was also one of the largest spacecraft ever sent into lunar orbit. Its mission was to be part of a duo of Radio Astronomy Exploration [RAE] missions that were to orbit the Moon and study low frequency emissions from the Sun, the surrounding planets and any other extragalactic sources it could find. *Explorer 50*, launched on 26 September 1973, was put into a geocentric orbit around the Earth. In June 1975 the programme was shut down, but contact was maintained with both Explorers until August 1977.

Skylab III

Launched on 28 July 1973 the second crew aboard *Skylab*, Alan L. Bean (commander), Jack R. Lousma and Owen K. Garriott, spent fifty-six days orbiting the earth. They carried out a variety of experiments and continued to maintain the space station. Among the experiments carried out was the S.150 X-ray experiment, the main purpose of which was to observe and record faint galactic X-ray sources. The experiment was housed in the upper section of the SIV-B stage and attached to the inside wall of the instrument unit that was housed there. When the spacecraft separated from the SIV-B stage the experiment was deployed and activated automatically. The SIV-B stage was then put through a series of maneuvers, which allowed the instruments, housed within the experiment, to scan pre-selected areas of the sky. The resulting data was recorded onto a tape recorder and broadcast to ground stations as it passed over them.

Other experiments included the study of the mass, speed and chemical composition of interplanetary dust and particles. A new type of Ultra Violet (UV) camera was also tested which enabled wide-field images to be taken and served as a photometer for stellar spectrography. A far-UV Electrographic camera was also used to study the structure of the Comet Kohoutek [named after Czech astronomer Lubos Kohoutek]. With these experiments completed and the housekeeping duties done, *Skylab III* returned to earth.

Skylab IV

The third and final crew, consisting of Gerald P. Carr (commander), William R. Pogue and Edward G. Gibson, launched from Cape Kennedy on 16 November 1973. They carried out the final selection of tests and proved conclusively that man could live and work in space in relative comfort. The *Skylab* crews carried out numerous experiments, including the study of the Sun during solar flares, the flux measurement of cosmic rays, the analysis of the solar corona, X-ray spectrography of solar flares and other active regions of the sun.

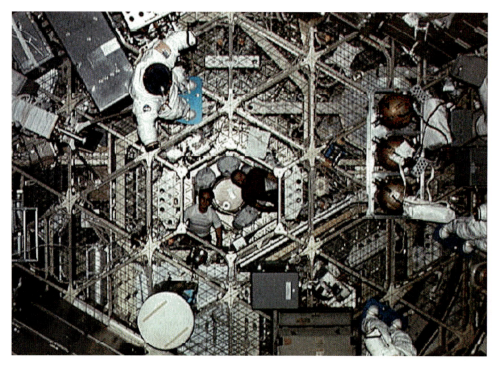

Inside *Skylab 4*. (NASA)

There were problems from day one when Bill Pogue suffered a serious bout of space sickness. The problems started when the other two crew members decided not to tell mission control, but they forgot that everything that was said on the space station could be heard by mission control. Because of Bill Pogue's reduced capacity to carry out assignments, the work list was falling behind. The mission had been extended from fifty-six days to eighty-four days and consequently the workload on the other two astronauts became increased. Then more medical experiments were factored into the extra days, causing the three astronauts to struggle to keep up with the frantic pace of the timeline. The crew were having no time for familiarisation or to recover from errors and malfunctions in some of the hardware. In between all this they had to carry out three spacewalks and housekeeping duties. A sixteen-hour day was becoming the norm and there were repeated requests from the crew for the workload to be reduced, but it fell on deaf ears. The *Skylab* flight director admitted later that they had got it wrong. Relations between the crew and mission control became very tense. During one orbit of the Earth the crew ceased communication with mission control, causing some concern. It soon became obvious that the workload was causing the crew to become fatigued, so it was decided to allow them a day of rest and reduce the workload. The *Skylab* programme was unfairly blighted by the *Skylab 4* incident and although rumours about the crew going on strike were mooted, the crew always denied this.

After the last crew had left *Skylab*, it was thought that the Shuttle would be able to dock with it and move it into a higher orbit, but solar flares from the sun caused the space station's slowly decaying orbit to accelerate and it was deemed to be redundant. *Skylab* was allowed to fall into a decaying orbit around the Earth and burnt up on re-entry on 11 July 1979, spreading remnants over the Indian Ocean and Western Australia. When the idea of a space laboratory was first put forward, there were a number of skeptics who had doubts about the usefulness of such a vehicle. Among these was the director of Kitt Peak Observatory, Leo Goldberg, who said at the end of the last mission:

> Many of us had serious doubts about the scientific usefulness of men in space, especially in a mission such as the ATM [Apollo Telescope Mount], which was part of the solar observatory, was not designed to take advantage of man's capability to repair and maintain equipment in space. But these men performed near-miracles in transforming the mission from near-ruin to total perfection. By their rigorous preparation and training and enthusiastic devotion to the scientific goals of the mission, they have proven the value of men in space as true scientific partners in space science research.

This statement summed up to perfection the feelings of the American scientist and reinforced NASA's commitment to the space programme, but other more reliable sources said that the mission to the ATM was a total failure. It appears that the automatic docking system failed and the mission had to be aborted, hence the mission only lasted two days. The crew of *Skylab IV* returned to Earth on 28 August 1973.

Although the American lunar programme had come to a halt, the Russians continued with the launch on 29 May 1974 of *Luna 22*. Like other Luna spacecraft, it continued to survey the lunar surface, its magnetic field, the composition of the lunar rocks and the surface Gamma ray emissions. The spacecraft went into a decaying orbit on 2 September when its maneuvering fuel was exhausted and it crashed on to the Moon's surface.

On 28 October 1974 the Russians launched *Luna 23*. This mission was to soft land the spacecraft and, by using its drill and scoop, acquire some samples and return them to Earth. Unfortunately on landing *Luna 23* tipped over, damaging the drill. Alternative plans to use the now defunct and stationary spacecraft for other experiments failed and the project was lost. Photographs taken by NASA's Lunar Reconnaissance Orbiter in 2012 showed *Luna 23* lying on its side.

CHAPTER ELEVEN

Apollo/Soyuz Test Project [ASTP]

T he 'space race', although it appeared to create a chasm between the United States and Russia, did in effect draw the two countries marginally closer. In 1962, just after John Glenn's historic spaceflight, Russian President Nikita Khrushchev sent President John F. Kennedy a telegram of congratulations, which included a statement about the need for co-operation between the two nations in the exploration of outer space. This resulted in a series of meetings between Deputy NASA Administrator Hugh Dryden and Soviet scientist Anatoly Blagonravov. What brought this about was the concern of both countries of the use of space around the Earth, and it was decided that some form of co-operation was needed. The launching of satellites in increasing numbers was causing some concerns, as if it continued unabated, it would result in a veritable 'junkyard' of decayed and decaying scrap metal to orbit the Earth. There was also the possibility of working satellites colliding with some of this debris and in the event of a manned space flight, danger to the lives of the astronauts. After the meetings, President John F. Kennedy proposed a joint world weather satellite system, using satellite observations and data exchange, thus limiting the number of satellites needed. He also proposed the sharing of spacecraft development and tracking and the mapping of the Earth's magnetic field. Later that year the two major powers signed an agreement on space co-operation with a number of the earlier proposals included.

Although they had signed the agreement, it initially only brought some limited response from the Russians, but after the *Apollo 13* incident it was realised that in the event of any spacecraft getting into difficulties, there should be some common ground that would enable either country to help set up a rescue mission. The Russians then put forward a number of proposals that the two countries shared experiences concerning the biological and medical experiments that had been carried out on the manned space flights. Initially it was only a small scale project, but by 1969 it had become much larger and more intensive.

One of the other proposals was for a joint space mission, this of course meant a sharing of technology which up to this point had been kept secret. The planning concerning a joint American/Russian space mission came a step closer on 2 December 1974, when cosmonauts Anatoliy Valis'yevich Filipchenko and Nikolay

Nikolayevich Rukavishnikov clambered aboard their *Soyuz 16* spacecraft, lifted off from Baikonur Cosmodrome and into space. This was the last of four test flights; the previous three were unmanned, and was to test the docking system by retracting and extending a simulated American designed and built docking ring. They also carried out tests on new solar panels, a new radar docking system, a modified environmental system and an improved control system. The docking ring was later jettisoned using explosive bolts in a simulated exercise to test the emergency measures in the event of the capture latches not releasing. *Soyuz 16* returned to Earth on 8 December 1974, landing near the town of Arkalvk. Their mission took six days and was later to be found to have been within ten minutes of the same mission time of the ASTP [Apollo Soyuz Test Program] itself. The whole mission was deemed to have been a great success. The scene was now set for a historic meeting between America and Russia, albeit in space.

To enable the two spacecraft to join up, a number of modifications had to be made to both capsules. The docking module, complete with a chamber through which the crews could pass, was built in the United States and had to be compatible with both spacecraft. The docking module was basically a cylindrical aluminum tube measuring 10 ft 4 in [3.15 m] long and 4 ft 8 in [1.4 m] in diameter. It contained television and communications equipment, together with gases to replenish the atmosphere, research apparatus and an electric furnace which was used by the Russians in a metal bonding experiment. There was considerable criticism of each other's spacecraft mainly because the Soyuz capsule was almost all automated and controlled from the ground, whereas the Apollo capsule had been designed to be flown by the crew. Christopher C. Kraft, director of the Manned Space Center, said:

> We in NASA rely on redundant components – if an instrument fails during flight, our crews switch to another in an attempt to continue the mission. Each Soyuz component, however, is designed for a specific function: if one fails, the cosmonauts land as soon as possible. The Apollo vehicle also relied on astronaut piloting to a much greater extent than did the Soyuz machine.

On 15 July 1975, *Soyuz 19* lifted off the launch pad at Baykonur, Kazakhstan and went into orbit. Then seven and a half hours later an Apollo spacecraft lifted off the launch pad at the Kennedy Space Center and went into orbit. The prime crew consisted of American astronauts, Tom Stafford, Vance DeVoe Brand and Donald 'Deke' Slayton. The back-up American crew were Alan Bean, Ron Evans and Jack Lousma. Jack Swigert had originally been assigned as the CM pilot, but had been removed after his alleged part in the *Apollo 15* commemorative postal cover scandal. The Russian crew were Colonel Aleksey Arkhipovich Leonov and civilian engineer Valeriy Nikolayevich Kubasov; the back-up Russian crew were Anatoly Filipchenko and Nikolai Rukavishnikov. Training for all the astronauts and cosmonauts had taken place both in America and in Russia and went well with both crews working together with no problems.

Apollo ASTP crew. (NASA)

Such was the interest in this historic meeting that the launch of the *Soyuz 19* spacecraft was broadcast live around the world, something that had never happened before in the history of the Russian space programme. In fact very few westerners have ever witnessed the launch of a Russian spacecraft or satellite. The two spacecraft, now in orbit, maneuvered into position and edged towards each other, then at a height of 140 miles above the Atlantic Ocean, they met. Minutes later the two spacecraft docked together and history was made. On the television screens in Mission Control, Houston, Texas and in Soyuz control at Kaliningrad, Russia, appeared the close-up of a hatch. Suddenly the hatch opened and the smiling face of Russian cosmonaut Alexi Leonov appeared, then a hand stretched out and the face of American astronaut Tom Stafford appeared on the screen. The two men shook hands warmly and greeted each other. The Apollo/Soyuz link-up cemented the friendship between the Russian and American astronaut/cosmonaut fraternity that had sprung up during training and also between Alexi Leonov and Tom Stafford. Later both crews got together and carried out the first ever in-space press conference.

For the next four days the two crews were in and out of each others' spacecraft and worked together to carry out a variety of experiments. One experiment carried out by the Russians consisted of three small cylinders that contained metals that could not be

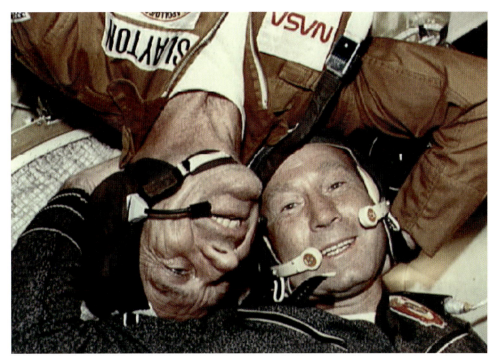

Tom Stafford and Alexi Leonov at the moment of their historic meeting in space. (NASA)

uniformly mixed on the ground because of the Earth's gravity. On Earth the heavier of the metals, in their molten state, settled to the bottom before the mixture cooled and solidified. The astronauts took the experiment, loaded the cylinders into an electric furnace in the docking module and proceeded with the sequence of heating, melting and cooling the samples in the weightless conditions. The test samples were returned to the cosmonauts for their scientists to evaluate the results. There were a number of medical experiments also carried out covering mainly biological investigations.

Amongst the experiments was a new kind of telescope that enabled the astronauts to look at the stars from above the Earth's atmosphere. Whilst they were searching for extreme ultraviolet radiation, they discovered an example way outside the solar system which enabled them to locate stars that up to this point scientists never knew existed. Among these stars was found a 'White Dwarf' star, made up of matter so heavy and dense that a small pebble could weigh a ton. This also led to the detection of X-ray emissions from stars, culminating in the discovery of the first known pulsar outside of the Milky Way galaxy. Situated approximately 200,000 light years away it was the most luminous pulsar ever found.

The number of independent experiments conducted jointly showed that the two nations could work together for common interest and set the benchmark for future space stations. One of the scientific experiments included an arranged eclipse of the Sun by placing the Apollo module in front of it to allow instruments on the Soyuz to take photographs of the solar corona.

Just prior to separation the two crews held a joint press conference where they answered questions from members of the press. The astronauts and cosmonauts were asked if there were any problems during the mission, to which the reply was a resounding: No! One of the 'tongue-in-cheek' comments came from a grinning Vance Brand, when he said that they were all so busy carrying out experiments in and out of the spacecraft that it needed a traffic cop to control and direct the flow of people. The two crews said their farewells and undocked, but half an hour later re-docked in an exercise to carry out a check on the workability of the docking mechanism, With that satisfactorily completed, the two spacecraft undocked for the last time. The Soyuz spacecraft went into orbit prior to re-entry, then two days later landed back in the USSR, and, like the launch, the whole episode was captured by hovering helicopters and shown on television, a first for both the Russian and American people. The Apollo spacecraft remained in orbit for a further five days in which the crew conducted more experiments.

There was almost a serious problem for the Apollo crew, when on re-entry, at 23,000 feet [7,010 m] and minutes before splashdown, the crew noticed yellow fumes inside the capsule. The crew suddenly experienced severe coughing and eye irritation, the gas was later identified as nitrogen tetroxide that had escaped from the RCS (Reaction Control System) thrusters.

On hitting the water, the spacecraft turned upside down with the nose under the water and the astronauts hanging face downwards in their seats, held only by their restraining harness. Stafford immediately released the flotation balloons and the spacecraft slowly righted itself. The atmosphere inside the spacecraft caused Vance Brand to lose consciousness and made the remaining two astronauts very woozy, but Stafford managed to reach oxygen masks from behind their seats and quickly ensured everyone had them on. Once the spacecraft was stabilised and in no fear of taking on seawater, Tom Stafford cracked the hatch to let fresh air in to the spacecraft. The crew quickly recovered and none of them suffered any ill effects from the experience. The crew stayed inside the spacecraft until it had been winched aboard the amphibious assault ship USS New Orleans [LPH-11].

When the two spacecraft rendezvoused in space they opened up another chapter in man's desire to explore the universe. The whole mission was a complete success for both technology and for the world. The American astronauts said afterwards, that the construction of the Russian spacecraft looked more like a plumber's nightmare than a sophisticated spacecraft. But what was also very apparent was that there was a true Esprit de Corps between the astronauts and the cosmonauts. It showed that East and West could work together in the furtherance of peace and harmony, as has since been proved through the Russian space station Mir and the ISS (International Space Station).

The American and Soviet commanders, Tom Stafford and Alexi Leonov, became lasting friends. Leonov became the godfather to Tom Stafford's younger children and Stafford gave a eulogy at Leonov's funeral in October 2019.

The Russian Luna programme ended on 9 August 1976 with the launch of Luna 24. After a soft landing on the lunar surface in the Mare Crisium region, the unmanned spacecraft collected 6 ounces [170 grams] of lunar samples and placed them in a sealed capsule. The spacecraft returned to Earth safely, landing in Siberia. The next spacecraft to visit the Moon was *Clementine*, launched on 25 January 1994. This was a project by the Americans to examine in detail the lunar surface, including both poles. In addition to this the spacecraft was to attempt to visit Asteroid 1620, but because of a computer problem which caused the spacecraft to burn up all of its fuel, the mission was cancelled. After completing all its tasks, *Clementine* was returned to Earth orbit.

It wasn't until 14 March 1995 before the next American, Norman Thagard, was to go into space, when the Russian *Soyuz TM-21* spacecraft delivered him to the Russian space station *Mir*.

CHAPTER TWELVE

The Next Generation

On 24 October 2007, the China National Space Administration [CNSA] started their Lunar Exploration Programme, known as Chang'e, by launching their first lunar spacecraft, *Chang'e 1*, to the Moon. The name Chang'e is that of the Chinese goddess of the Moon. The programme incorporated lunar orbiters, landers, rovers and sample-returning spacecraft. The ultimate aim is to send a manned spacecraft to land on the Moon, but nothing is scheduled for the near future. The launch vehicle was a Long March 3A rocket and was launched from Xichang Satellite Launch Centre in Sichuan Province in the south-west of China. This is one of three spaceports, the others are at Taiyuan in the north and Jiuquan in the north-west. Jiuquan has recently been expanded to facilitate both launches of commercial solid rockets and new methane-liquid oxygen rockets.

After a twelve-day journey, *Chang'e 1* orbited the Moon carrying out one of the most detailed photograph mapping missions to date, looking for possible landing sites. It also mapped and identified the various chemical elements that make up the lunar surface, as part of an evaluation for possible useful resources.

Japan joined the elite Lunar Space Club on 22 September 2007 when they launched their spacecraft *Selene*, also known as *Kaguya*, from the Tanegashima Space Center on an H-11A carrier rocket. This was the second mission planned; the first had been scheduled to launch in 2003 but had been cancelled because of electronics problems.

The *Kaguya* spacecraft entered into lunar orbit on 3 October and released a relay satellite over the lunar south pole. At the end of October the *Kaguya* spacecraft deployed a Lunar Magnetometer, a Lunar Radar Sounder and an Earth-looking Upper Atmosphere and Plasma Imager into orbit. Over the next two months all three pieces of equipment were tested before, on 21 December, all fifteen observation experiments were deemed to be up and running. At the end of October 2008, with all the experiments successfully completed, *Kaguya* was placed in a gradual decaying orbit around the Moon before finally crashing into the lunar surface on 10 June 2009.

In October 2008 India became the fifth country to visit the Moon, when they launched their spacecraft *Chandrayaan-1*. It was launched from the Satish Dhawan Space Center at Srihankota, Andhra Pradesh, South India, aboard a PSLV-XL rocket. On 8 November the spacecraft went into lunar orbit where it orbited the Moon for six days before landing at the south pole. It was intended to operate for two years, surveying the lunar surface and carrying out a complete chemical analysis of the surface in addition to producing a three-dimensional topography. One of its achievements was the discovery of a widespread presence of water molecules in the lunar soil. Less than one year into its mission, the spacecraft started to suffer a number of technical problems, firstly with the thermal shields and then the star tracker. Then on 28 August 2009 all communications with the spacecraft stopped and the mission was declared over. However almost all the assigned projects were completed, so despite having to shut down early, the mission was declared to have been a success.

Three years after *Chang'e 1* was launched, the Chinese Space Agency launched *Chang'e 2* on 7 October 2010 aboard a Long March 3C rocket, and after just a five-day journey, reached the Moon. After carrying out another detailed mapping of the Moon, the spacecraft headed towards the sun to test the telemetry and tracking system. After completing this part of the mission, it then carried out a fly-by of Asteroid 4179 before heading off into deep space. It was never destined to be recovered, but served to see how long they were able to track the spacecraft before it disappeared into the never-ending blackness of space.

Chang'e 3 was launched 2 December 2013 on a Long March 3B rocket. This was to be China's first lunar landing and on board the spacecraft was the lunar rover *Yutu-1*. After landing in the *Mare Imbium* area it carried out a number of experiments and surveyed an area covering three square miles. The mission was a great success and opened the way for the next flight to the Moon, *Chang'e 4*, which was launched on 7 December 2018. *Chang'e 4* had originally been scheduled as a back-up for *Chang'e 3*, but with the success of that mission the configuration was changed. Launched on 7 December 2018 on a Long March 3B rocket, the spacecraft landed on the far side of the Moon in the *Von Karman Crater* in the South Pole-Aitken Basin, where it unloaded the second of China's lunar rovers, *Yutu 2*. The lunar rover explored the area around the Pole-Aitken Basin collecting samples, which were later returned to Earth. A terrain camera that was able to rotate 360° was mounted on top of the lander and captured a large number of very detailed photographs.

Israel joined the space club on 21 February, when their unmanned spacecraft *Beresheet-1* was launched on a Space X Falcon 9 rocket from Cape Canaveral, Florida. After a long uneventful space flight, the spacecraft went into orbit around the Moon. On 11 April whilst attempting a soft landing, it suffered a malfunction, crashed onto the lunar surface and was lost.

On 22 July 2019, boosted with the successful landing on the Moon of *Chandrayaan-1*, *Chandrayaan-2* was launched aboard a PSLV-XL rocket from the Satish Dhawan Space Center at Srihankota, Andhra Pradesh, South India. The mission was to follow up on the previous mission at the lunar south polar region, searching for water deposits. On board was a lunar rover that was to enable scientists to obtain samples from various areas of the South Pole region. On reaching the Moon on 22 August, the spacecraft went into a lunar orbit until 6 September when its lander started to make its descent onto the lunar surface. As it approached the surface for a soft landing,

Above: Chinese lunar rover *Yutu 2*. (CNSA)

Below: Chinese lander spacecraft on the Moon. (CNSA)

it suddenly deviated from its intended course and crashed heavily onto the surface and was lost. Despite numerous attempts to make contact with the spacecraft, none were successful and the decision was taken to abandon the mission. However the *Chandrayaan-2* orbiter continued to orbit the Moon sending back information and was later to become an important part of of India's next mission, *Chandrayaan-3*.

The CNSA launched *Chang'e 5* on 25 October 2020 aboard a Long March 5 rocket and, after landing close to Mons Rümker, collected more than 4.4 lb [2 kg] of lunar samples and returned them to Earth. This was to be the last Chinese mission to the Moon for some time as the next, *Chang'e 6*, is not scheduled to be launched until 2025.

In 2017, under the Trump administration, the Americans announced that they were going back to the Moon and formally established the Artemis programme. Artemis was the Greek goddess of the Moon and the twin sister of Apollo, giving a link back to the last mission to the Moon – *Apollo 17*. The concept initially had its roots back in 2005 under the George W. Bush administration, with the Constellation programme and the development of the Orion spacecraft, but during the Obama administration the programme was cancelled. In 2016 it was re-awakened and the launch of the Orion spacecraft, using the Space Launch System, was scheduled that same year. Once again the programme was placed on hold and re-scheduled for launch on 16 November 2022. Known as *Artemis 1*, it was manned with robots and three life-size mannequins aboard all dressed in Orion Crew Survival System spacesuits. After splashing down in the Pacific Ocean on 11 December 2022, the spacecraft was recovered by the *USS Portland* [LPD-27] and taken to the US Naval base in San Diego. It was then taken by road to the Kennedy Space Center in Florida. With the success of this first flight, a fully crewed Orion spacecraft, *Artemis 2*, with four crew members on board, is scheduled for launch in November 2024 to carry out a fly-by of the Moon.

The Orion spacecraft was launched using the Space Launch System [SLS] which is the most powerful rocket in the world today and is powered by four RS-25 engines and twin five-segment solid rocket boosters. Once in orbit the Interim Cryogenic Propulsion Stage [ICPS] will send the spacecraft onto the Moon. For future flights it is planned to have an extraterrestrial space station, known as the Lunar Gateway, in lunar orbit, which will serve as a staging post for both manned and unmanned exploration of the Moon and later for the manned exploration of Mars and beyond.

A Japanese private enterprise firm joined with the Japanese Space Agency and carried out an unmanned lunar mission with their spacecraft *Hakuto-R*. Launched on 11 December 2022 on a Falcon 9 rocket from Cape Canaveral, the spacecraft ran out of fuel as it made its descent onto the lunar surface and crashed, destroying it completely.

On 10 August 2023 an unmanned Soyuz 2.1b rocket was launched from the Vostochny Cosmodrome in the Amur region of East Russia, carrying *Luna 25* to the Moon. The previous Lunar mission was *Luna 24*, which was launched almost fifty years ago and successfully returned to Earth with samples of Lunar material. *Luna 25* was scheduled to stay on the lunar surface for at least a year, during which time it would collect and analyse samples and send the results back to Earth. The cameras that were installed on the spacecraft have already transmitted a number of images of both the lunar surface and of the Earth whilst in lunar orbit. Unfortunately, whilst making its descent onto the Moon's South Pole, the spacecraft suffered a problem within its computer guidance system, causing it to crash onto the lunar surface. All communication with the spacecraft was lost and it was deemed to have been destroyed.

The war in Ukraine is now having a detrimental effect on the Russian Space programme, as a number of countries have said that they will not co-operate with Russia whilst this conflict goes on. This of course creates a major problem with the International Space Station [ISS] as Russia is one of the main transporters of crews and supplies.

India continued with its space programme, after watching with great trepidation, the loss of Russia's *Luna 25* when it attempted to land close to the South Pole. They launched their unmanned *Chandrayaan-3* on an LVM3 rocket, from the Satish Dhawan Space Center at Srihanikota, Andhra Pradesh, south India on 14 July 2023. The Lander spacecraft *Vikram* carried out a soft landing on the area close to the lunar South Pole on 24 August 2023. India became the first country to carry out a soft landing in this area and the fourth to successfully soft land on the lunar surface. Carried in the Lander *Vikram* is a six-wheeled, solar powered lunar rover, *Pragyan* [Wisdom]. All communication from the rover is made through the lander, which in turn will transmit the information to the *Chandrayaan-2* orbiter, which is still orbiting the Moon, and from there back to Earth. The rover will leave the Lander and be active for one lunar day [29.5 days on Earth]. During this time it will collect and analyse lunar soil samples and, by using an Alpha Particle X-ray Spectrometer [APXS], will look for elements in the lunar soil and rocks. The main mission is to search for the presence of water, possibly in the form of ice, especially in the dark craters of the South Pole area. These permanently shadowed regions have temperatures as low as -418 degrees Fahrenheit [-250 degrees Celsius], which incidentally is colder than the planet Pluto. As we all know, water is essential for life, but it has many other uses; it can act as a coolant for equipment, provide drinking water and can even be used to provide fuel. In the event of a base being established on the Moon, the water could be broken down into hydrogen and oxygen to provide breathable air. From a purely scientific basis it can be used to detect and record geological activity and track asteroid strikes. It has been asked, if the South Pole is deemed to be so important, why is it only now that they have managed to land on it, considering the number of unmanned and manned missions that have been made to the Moon? It appears that because the South Pole is almost permanently in shadow, it is extremely hazardous to land in the area, but it is also the one place where it is thought there is probably an abundance of ice. Time will tell if these landings prove to be advantageous to future visitors. *Artemis* is scheduled to become the next mission, in 2024, followed by the first manned mission in 2025, fifty-three years after *Apollo 17*.

Although NASA and the ESA are the prime movers in *Artemis*, it is in fact a collaboration of different governments and their space agencies, together with private spaceflight companies. Along with NASA and the ESA, twenty-three countries have signed the Artemis Accord, including the United Kingdom, Canada, Japan, Brazil, South Korea and the United Arab Emirates.

The *Artemis 2* crew that will be the first to fly to the Moon for more than fifty years includes the first non-American, Canadian Jeremy Hansen. The rest of the crew are Reid Wiseman [Commander], Victor Glover [Pilot] and Christina Koch [Mission Specialist].

However there were further attempts to reach the Moon and on 22 February 2019, Israel, together with SpaceIL, launched a Lunar Lander on a Falcon 9 B5 rocket called Beresheet, with the intention of carrying out a soft landing on the surface of the Moon. The lander's gyroscopes failed on 11 April 2019 causing the main engine to shut down, which resulted in the lander crashing on the Moon's surface and was lost. This was followed three years later by another attempt, Peregrine 1. On the 8 January 2024, a private consortium launched a Vulcan Centaur rocket from Cape Canaveral, Florida called Peregrine 1, to carry out an unmanned mission to land on the Moon. After a flawless launch and first stage separation from the Vulcan Centaur rocket, problems arose with the power supply. The solar panels, which powered the batteries, would not deploy properly, then a stuck valve caused a significant leak in the fuel tank which depleted the propellant. Without the fuel the spacecraft would be unable to carry out mid-course corrections and manoeuvre into position to land on the surface of the Moon. The spacecraft re-entered the Earth's atmosphere and was burnt up. Twelve days later, on 20 January 2024, the Japanese spacecraft SLIM was launched and successfully carried out a landing on the surface of the Moon despite suffering an engine problem in the last few minutes before touchdown. The lander, known as the 'Moon Sniper' because of its pin-point accuracy, landed within feet of its intended landing area albeit at a skewed angle and almost upside down. Unfortunately, a problem then arose when it was discovered that its solar arrays were not generating enough power to support the batteries. The lander was then switched off until the sunlight reached the solar panels and started regenerating the solar batteries. Five days later the lander was re-activated, but because of its skewed angle its mission has become very limited. It will continue to observe and collect information, but any thoughts of trying to right the lander have been dismissed, as it might lead to placing it in a more disastrous position. Just one month later, on 15 February 2024 a commercial spacecraft called Nova C-Odysseus was successfully launched from Cape Canaveral, Florida, on top of a Space X Falcon 9 rocket. The spacecraft successfully landed on Malpert A, close to the Moon's south pole. This was the first US spacecraft to land on the Moon since Apollo 17 over fifty years ago. Unfortunately NASA, who had been monitoring the mission, announced that the spacecraft had tipped over and was lying on its side.

One hundred and twenty-one years ago the Wright Brothers launched the first powered flight, and in that relatively short period of time, man has achieved the seemingly impossible: he has walked on another world and sent robotic probes to far distant planets. It will be interesting to see what man will achieve in the next hundred years or so.

Lunar Spaceflight Chronology

The following is a chronology relating to spaceflights to the Moon:

Luna 1. [Russia] Unmanned. Launched on 2 January 1959 with the intention of crashing onto the Moon, but spacecraft flew by.

Luna 2. [Russia] Unmanned. Launched 12 September 1959, it became the first spacecraft to land on the Moon.

Luna 3. [Russia] Unmanned. Launched 4 October 1959, Luna 3 was the first to orbit the Moon and take photographs of the far side.

Mercury-Redstone I. [USA] Unmanned. Launched on 21 November 1960, the rocket shut down just after launch and jettisoned its escape rocket.

Mercury-Redstone II. [USA] Launched on 31 January 1961, this was the first flight of a primate, Ham, who carried out a sub-orbital flight.

Vostok 1. [Russia] Launched on 12 April 1961, this was the first manned spaceflight, made by Yuri Gagarin who orbited the Earth just the once.

Mercury-Redstone III. [USA] Launched 5 May 1961, the first American in space, Alan Shepard.

Mercury-Redstone IV. [USA] Launched 21 July 1961. This was the second manned sub-orbital flight, with Gus Grissom aboard.

Vostok 2. [Russia] Launched 6 August 1961. Cosmonaut Gherman Titov orbited the Earth seventeen times in the first prolonged period in space.

Ranger 1. [USA] Launched 23 August 1961. Unmanned engineering test flight that was supposed to enter an elliptical orbit that would take it beyond the Moon, but a malfunctioning rocket placed it in a low Earth orbit where it finally re-entered the Earth's atmosphere and burnt up.

Ranger 2. [USA] Unmanned engineering test flight launched 18 September 1961. Like *Ranger 1* there were problems which resulted in the spacecraft being placed in an even lower Earth orbit and it too burnt up on re-entry.

Mercury-Atlas V. [USA] Launched 29 November 1961, this was the first of the primates, Enos, from the US to orbit the Earth.

Ranger 3. [USA] Launched 26 January 1962. Equipped with a television camera and a seismometer, the mission was to land the spacecraft on the Moon, but a computer failure caused it to miss the Moon and skip it to space.

Mercury-Atlas VI. [USA] Launched 20 February 1962. Astronaut John Glenn was to be the first American to make a manned orbital spaceflight, completing three orbits of the Earth.

Ranger 4. [USA] Launched 23 April 1962. The unmanned spacecraft was programmed to take photographs of the Moon and then make a hard landing on the surface. A computer problem caused everything to shut down and the spacecraft to crash on the far side of the Moon.

Mercury-Atlas VII. [USA] Launched on 24 May 1962. Scott Carpenter was the second American astronaut to orbit the Earth.

Vostok 3. [Russia] Launched 11 August 1962. Cosmonaut Andrian Nikolayev completed four orbits.

Vostok 4. [Russia] Launched 12 August 1962. Cosmonaut Pavel Popovich completed forty-eight orbits of the Earth. This and *Vostok 3* got within five kilometres of each other.

Mercury-Atlas VIII. [USA] Launched on 3 October 1962 with Wally Schirra. This flight was to test the technical side of the spacecraft.

Ranger 5. [USA] Unmanned, launched 18 October 1962. The mission was to make a hard landing on the Moon, but the batteries drained prematurely and the spacecraft missed its target and went into a heliocentric orbit.

Luna 4. [Russia] Unmanned. Launched 2 April 1963, it was placed initially in an Earth orbit and then sent to the Moon. A computer hitch caused the spacecraft to miss the Moon and go into a heliocentric orbit.

Mercury-Atlas IX. [USA] Launched 15 May 1963, this was the last of the Mercury missions, and in which Gordon Cooper carried out 22 orbits of the Earth and spent thirty-four hours in space.

Vostok 5. [Russia] Launched 14 June 1965. Cosmonaut Valeriy Bykovskiy was to meet up with *Vostok 6*. Maintained radio contact but no visual.

Vostok 6. [Russia] Launched 16 June 1963. Cosmonaut Valentina Tereshkova was the first woman to go into space and was to join up with *Vostok 5*.

Ranger 6. [USA] Unmanned. Launched 30 January 1964. This was the first of the Ranger probes to have six television cameras with the intention of televising the hard landing on the lunar surface. A power problem caused them all to malfunction.

Gemini I. [USA] Launched 8 April 1964. Unmanned spacecraft to test the structure. Thirty-three orbits carried out.

Ranger 7. [USA] Unmanned. Launched 28 July 1964. This was the first probe to successfully send close images of the lunar surface back to Earth, via the six television cameras mounted on it. It then carried out a successful hard landing.

Voskhod 1. [Russia] Launched 12 October 1964. This was the first multi-manned spacecraft to go into orbit around the Earth. Crew: Vladimir Komarov, Konstantin Feoktistov, and Boris Yegorov.

Gemini II. [USA] Launched 19 January 1965. This was the second unmanned mission to test the spacecraft.

Ranger 8. [USA] Unmanned. Launched on 17 February 1965, this was the most successful mission to date, taking more than 7,000 close-up photographs of the lunar surface.

Voskhod 2. [Russia] Launched 18 March 1965. The second multi-manned spaceflight, crewed by Pavel Belyayev and Alexi Leonov, and the first spacewalk, carried out by Alexi Leonov.

Ranger 9. [USA] Unmanned. Launched 21 March 1965 this mission was a complete success, very similar to *Ranger 8*'s mission.

Gemini III. [USA] Launched 23 March 1965, this was the first two-manned American spaceflight, with Gus Grissom and John Young.

Luna 5. [Russia] Unmanned. Launched 9 May 1965. Crashed on the Moon after gyroscopic failure.

Gemini IV. [USA] Second of the manned Gemini missions, with Jim McDivitt and Ed White.

Luna 6. [Russia] Unmanned. Launched 8 June 1965 for intended soft landing on the Moon but due to a failed mid-course correction missed.

Zond 3. [Russia] Unmanned. Launched 18 July 1965. Took photographs of the far side of the Moon before going into a heliocentric orbit. *Zond 1* and *2* were Venus and Mars flybys.

Gemini V. [USA] Launched 21 August 1965. The crew of Gordon Cooper and Pete Conrad were in space for eight days.

Luna 7. [Russia] Unmanned. Launched 4 October 1965. Crashed onto the lunar surface and was lost.

Luna 8. [Russia] Unmanned. Launched 3 December 1965. Crashed onto the lunar surface and was destroyed.

Gemini VII. [USA] Launched 4 December 1965 to successfully rendezvous with *Gemini 6A*. Crew: Frank Borman and James Lovell.

Gemini VI-A. [USA] Launched 15 December 1967. The two-man crew Wally Schirra and Tom Stafford rendezvoused with *Gemini 7*.

Luna 9. [Russia] Unmanned. Launched 31 January 1966. This was the first Luna spacecraft to achieve a soft landing on the Moon and transmit photographs back to Earth.

Gemini VIII. [USA] Launched 16 March 1966. Crew: Neil Armstrong and David Scott. Carried out the first docking with another spacecraft.

Luna 10. [Russia] Unmanned. Launched 31 March 1966. First Russian spacecraft to orbit the Moon.

Surveyor 1. [USA] Unmanned. Launched 30 May 1966. First American soft-landing on the lunar surface.

Gemini IX-A. [USA] Launched 3 June 1966. Crew: Eugene Cernan and Tom Stafford. Unsuccessful link-up with GATV.

Explorer 33. [USA] Unmanned. Launched 1 July 1966 to study the Earth from a lunar distance.

Gemini X. [USA] Launched 18 July 1966. Crew: John Young and Michael Collins. Successful rendezvous with GATV.

Orbiter 1. [USA] Unmanned. Launched 10 August 1966. Orbited the lunar surface, sending back numerous photographs. Later crashed onto the surface of the Moon.

Lunar 11. [Russia] Unmanned. Launched 24 August 1966. Orbited the Moon taking photographs. Later crashed onto the lunar surface.

Gemini XI. [USA] Launched 12 September 1966. Crew: Charles 'Pete' Conrad and Richard Gordon. Performed the first one orbit rendezvous with ATV.

Surveyor 2. [USA] Unmanned. Launched 20 September 1966. Thruster problem caused the spacecraft to crash into the lunar surface.

Luna 12. [Russia] Unmanned. Launched 22 September 1966. Orbited the Moon carrying out scientific observations. Crashed into the lunar service one year later.

Orbiter 2. [USA] Unmanned. Launched 6 November 1966. Orbited the Moon photographing possible landing sites. Impacted the lunar surface one month later.

Gemini XII [USA] Launched 12 November 1966. Crew: James Lovell and Buzz Aldrin. Spacecraft docked with the GATV. This was the final flight of the Gemini programme.

Luna 13. [Russia] Unmanned. Launched 21 December 1966. Successful hard landing on lunar surface.

Apollo 1. [USA] Crew: Gus Grissom, Edward White and Roger Chaffee. Fire in the capsule on 27 January 1967 whilst testing killed the entire crew.

Orbiter 3. [USA] Unmanned. Launched 5 February 1967. Surveyed possible landing sites.

Surveyor 3. [USA] Unmanned. Launched 17 April 1967. First mission to carry a lunar surface scoop.

Orbiter 4. [USA] Unmanned. Launched 4 May 1967. Orbited the Moon, mapping the lunar surface.

Explorer 34. [USA] Unmanned. Launched 24 May 1967. Studied solar plasma radiation whilst orbiting the Moon.

Surveyor 4. [USA] Unmanned. Launched 14 July 1967. Crashed on landing and was destroyed.

Explorer 35. [USA] Unmanned. Launched 19 July 1967. Designed to study the magnetic field, solar X-rays, and interplanetary plasma whilst orbiting the Moon.

Orbiter 5. [USA] Unmanned. Launched 1 August 1967 to map the Moon for suitable landing sites for the Apollo missions. This was the last of the Orbiter missions.

Surveyor 5. [USA] Unmanned. Launched 8 September 1967. This was the first spacecraft to carry out a lunar soil analysis.

Apollo 4. [USA] Unmanned. 9 November 1967. Testing of the main and Lunar Module ascent and descent engine. First test of the Saturn V launch vehicle.

Surveyor 7. [USA] Unmanned. Launched 7 January 1968. The last of the Surveyor missions to explore the surface of the Moon. A total of 21,091 photographs were transmitted back to Earth during the Surveyor missions.

Apollo 5. [USA] Unmanned. Launched 22 January 1968. First test flight of the Lunar Module.

Zond 4. [Russia] Unmanned. Launched 2 March 1968. Part of the Soyuz programme and the first experiment for a manned lunar spacecraft.

Apollo 6. [USA] Unmanned. Launched 4 April 1968. The last of the unmanned Apollo test flights. Second test of the Saturn V launch vehicle.

Luna 14. [Russia] Unmanned. Launched 7 April 1968 to test communications from lunar orbit to ground stations on Earth.

Zond 5. [Russia]. Unmanned. Launched 14 September 1968. This was the first spacecraft to fly to the Moon, carry out an orbit and safely return to Earth. Two live tortoises were on the spacecraft, both survived.

Apollo 7. [USA] Launched 11 October 1968. Crew: Wally Schirra, Walter Cunningham and Don Eisele. The first manned flight of the Apollo programme.

Zond 6. [Russia] Unmanned. Launched 10 November 1968. Carried out a photographic orbital of the Moon and returned safely.

Apollo 8. [USA] Launched 21 December 1968. Crew: Frank Borman, James Lovell and William Anders. First crew to orbit the Moon [10 times] and return safely.

Apollo 9. [USA] Launched 3 March 1969. Crew: James McDivitt, Dave Scott and Russell Schweickart. First transposition of LM and docking with the CSM whilst in Earth orbit.

Apollo 10. [USA] Launched 18 May 1969. Crew: Thomas Stafford, John Young and Eugene Cernan. Full dress rehearsal for the first Moon landing.

Luna 15. [Russia] Unmanned. Launched 13 July 1969. Attempted to land on the lunar surface but crashed.

Apollo 11. [USA] Launched 16 July 1969. Crew: Neil Armstrong, Edwin 'Buzz' Aldrin and Michael Collins. First humans to set foot on another world.

Zond 7. [Russia] Unmanned. Launched 8 August 1969. Carried out photographic orbits of the Moon and returned safely to Earth.

Apollo 12. [USA] Launched 14 November 1969. Crew: Charles 'Pete' Conrad, Richard Gordon and Alan Bean. Second successful landing on the Moon.

Apollo 13. [USA] Launched 11 April 1970. Crew: James Lovell, Fred Haise and Jack Swigert. Explosion in SM caused Moon landing to be aborted. One of the most dramatic rescues in history. Crew returned safely.

Luna 16. [Russia] Unmanned. Launched 12 September 1970. First to land a robotic on the lunar surface and return to Earth with a sample.

Zond 8. [Russia] Unmanned. Launched 20 October 1970. Orbited the Moon, sending back photographs of possible landing sites. Originally meant to carry a cosmonaut crew member but cancelled at the last minute after it was thought that the safety measures were inadequate. Spacecraft returned safely, landing in the Indian Ocean.

Luna 17. [Russia] Unmanned. Launched 19 November 1970. Landed on the Moon and deployed the first lunar rover – *Lunokhod 1.*

Apollo 14. [USA] Launched 31 January 1971. Crew: Alan Shepard, Stuart Roosa and Edgar Mitchell. Initial difficulties in docking with LM almost caused the mission to be abandoned. Successful mission.

Apollo 15. [USA] Launched 26 July 1971. Crew: Dave Scott, Jim Irwin and Alfred Worden. First deployment of the Lunar Rover Vehicle [LVR] which was used throughout the mission.

Luna 18. [Russia] Unmanned. Launched 2 September 1971. Intended to bring back a lunar sample, but crashed on lunar surface.

Luna 19. [Russia] Unmanned. Launched 28 September 1971. Orbited the Moon for mapping mission.

Luna 20. [Russia] Unmanned. Launched 14 February 1972. Mission to obtain lunar sample successful, but return rough landing damaged sample.

Apollo 16. [USA] Launched 16 April 1972. Crew: John Young, Thomas Mattingly and Charles Duke. Successful fifth lunar landing mission .

Apollo 17. [USA] Launched 2 December 1972. The last manned flight to the Moon. Crew: Gene Cernan, Harrison Schmitt and Ronald Evans.

Luna 21. [Russia] Unmanned. Launched 8 January 1973. The second deployment of a *Lunokhod.*

Explorer 49. [USA] Unmanned. Launched 10 June 1973. Orbited the Moon to carry out experiments in radio astronomy.

Luna 22. [Russia] Unmanned. Launched 29 May 1974. Orbited the Moon, continuing to map the lunar landscape.

Luna 23. [Russia] Unmanned. Launched 28 October 1974. Landed on Moon to collect lunar sample and return it to Earth.

Luna 24. [Russia] Unmanned. Launched 9 August 1976. Landed on the Moon to collect lunar sample and return to Earth.

Prospector. [USA] Unmanned. Launched 7 January 1998. Orbited the Moon for nineteen months, mapping the polar regions in an effort to find polar ice.

Hakuto-R M1. [Japan] Unmanned. Launched 11 December 2022 on a Falcon 9 rocket from Cape Canaveral. Spacecraft ran out of fuel and crashed onto the lunar surface and was destroyed.

Luna 25. [Russia] Unmanned. The first Lunar mission since *Luna 24* in 1976. Launched on 10 August 2023 from Vostochny Cosmodrome, the unmanned *Luna 25* was scheduled to spend a year on the Lunar surface collecting and testing samples. Unfortunately the spacecraft was lost when it crashed as it attempted to land on the lunar surface.

Chandrayaan-3. [India] Unmanned. Launched 14 July 2023 on a PSLV-XL rocket from the Satish Dhawan Space Center at Srihankota, Andhra Pradesh, South India. Landed on the lunar South Pole on 24 August 2023. The first country to soft land in this area of the Moon.

Beresheet. On 22 February 2019, Israel, together with SpaceIL, launched a Lunar Lander on a Falcon 9 B5 rocket called Beresheet, with the intention of carrying out a soft landing on the surface of the Moon. The lander's gyroscopes failed on 11 April 2019 causing the main engine to shut down, which resulted in the lander crashing on the Moon's surface and was lost.

Peregrine 1. On the 8 January 2024, a private consortium launched a Vulcan Centaur rocket from Cape Canaveral, Florida called Peregrine 1, to carry out an unmanned mission to land on the Moon. After a flawless launch and first stage separation from the Vulcan Centaur rocket, problems arose with the power supply. Without the fuel the spacecraft was unable to carry out mid-course corrections and manoeuvre into position to land on the surface of the Moon. The spacecraft re-entered the Earth's atmosphere and was burnt up,

SLIM. On 20 January 2024, the Japanese spacecraft SLIM, successfully carried out a landing on the surface of the Moon despite suffering an engine problem in the last few minutes before touchdown. The lander, known as the 'Moon Sniper' because of its pin-point accuracy, landed within feet of its intended landing area albeit at a skewed angle and almost upside down. Five days later the lander was re-activated, but because of its skewed angle its mission has become very limited. It will continue to observe and collect information, but any thoughts of trying to right the lander have been dismissed, as it might lead to placing it in a more disastrous position.

Nova C. Odysseus. On 15 February 2024 a commercial spacecraft called Nova C -Odysseus was successfully launched from Cape Canaveral, Florida, on top of a Space X Falcon 9 rocket. The spacecraft successfully landed on Malpert A, close to the Moon's south pole. This was the first US spacecraft to land on the Moon since Apollo 17 over fifty years ago. It was discovered later that the spacecraft had tipped over and was now lying on its side.

Rocket Specifications

Little Joe

Height	55 ft [15.2 m]
Diameter	6 ft [2.03 m]
Mass weight	3,000 lb [1,400 kg]
Stages	2
Fuel	Solid propellants

Jupiter C 1

Height	69.9 ft [21.3 m]
Diameter	5.10 ft [1.8 m]
Mass weight	64,070 lb [29,080 kg]
Stages	3
Fuel	LOX and Hydyne

Juno 1

Height	69.9 ft [21.3 m]
Diameter	5.10 ft [1.8 m]
Mass weight	64,070 lb [29,080 kg]
Stages	4
Fuel	LOX and Hydyne

Vanguard

Height	72 ft [21.94 m]
Diameter	3.9 ft [1.14 m]
Mass weight	22,100 lb [10,050 kg]
Stages	3
Fuel	LOX and Kerosene

Mercury-Redstone

Height	83.38 ft [25.41 m]
Diameter	5.83 ft [1.78 m]
Mass weight	66,000 lb [30,000 kg]
Stages	Single
Fuel	Liquid Oxygen [LOX] and Alcohol

Mercury-Atlas

Height	94 ft [28.6 m]
Diameter	10 ft [3 m]
Mass weight	260,000 lb [117,934 kg]
Stages	Single
Fuel	Liquid Oxygen [LOX] and RP-1 kerosene

Gemini Missions

Titan II

Height	103 ft [31.3 m]
Diameter	10 ft [3.5 m]
Mass weight	340,000 lb [154,000 kg]
Stages	2
Fuel	Aerozine 50 and nitrogen tetroxide oxidiser

Apollo Missions

Saturn IB for all orbital flights.

Height	141.6 ft [43.2 m]
Diameter	21.7 ft [6.61 m]
Mass weight	1,300,220 lb [589,770 kg]
Stages	2
Fuel	Liquid Hydrogen [LH-2] and LOX

Saturn V for all Lunar Missions.

Height	363 ft [110.6 m]
Diameter	33 ft [10.1 m]
Mass weight	6.2 million lb [2.8 million kg]
Stages	3
Fuel	LOX and RP-1 [First Stage]

LH and LOX [Second Stage]
LH and LOX [Third Stage]

Sources

Personal correspondence with astronauts Jim McDivitt [*Gemini 1V, Apollo 9*], Walt Cunningham [*Apollo 7*], Bill Pogue [*Skylab*], Alan Bean [*Apollo 12, Skylab*], Al Worden and Jim Irwin [*Apollo 15*].

Bibliography

The All American Boys – Walter Cunningham. McMillan Publishing, New York – 1977.

Soviet Rocketry – Michael Stoiko. David & Charles, Newton Abbot, UK – 1971.

Russians in Space – Evgeny Riabchikov. Novosti Press, Moscow – 1971.

Apollo Soyuz – NASA Publications.

Return to Earth – Buzz Aldrin.

Mercury Project Summary – NASA Publications.

Gemini – NASA Publications.

Apollo – NASA Publications.

Exploring the Moon – David Harland, Praxis Publishing, Chichester – 1999.

Glossary

ABMA	Army Ballistic Missile Agency.
ALSEP	Apollo Lunar Surface Experiments Package.
ASTP	Apollo Soyuz Test Project.
ATDA	Agena Target Docking Adapter.
ATM	Apollo Telescope Mount.
CapCom	Capsule Communicator.
CCB	Configuration Control Board.
CIA	Central Intelligence Agency.
CM	Command Module.
CSM	Command and Service Module.
DoD	Department of Defence.
ESA	European Space Agency.
EVA	Extra Vehicular Activity.
GDL	Gas Dynamics Laboratory.
ISS	International Space Station.
JPL	Jet Propulsion Laboratory.
LES	Launch Escape System.
LH	Liquid Hydrogen.
LM	Lunar Module.
LLRV	Lunar Landing Research Vehicle.
LOX	Liquid Oxygen.
LTA	Lunar Module Test Article.
LRV	Lunar Roving Vehicle.
MET	Mobile Equipment Transporter.
MOL	Manned Orbiting Laboratory.
MQF	Mobile Quarantine Facility.
NASA	National Air and Space Administration.
NRL	Naval Research Laboratory.
PLSS	Portable Life Support System.
RCS	Reaction Control Thrusters.
SECO	Sustainer Engine Cut Off.
SM	Service Module.
UCD	Urine Collection Device.
UV	Ultra Violet.